Maximizing Value with Automation and Digital Transformation

Leslie P. Willcocks · John Hindle ·
Matt Stanton · John Smith

Maximizing Value with Automation and Digital Transformation

A Realist's Guide

Leslie P. Willcocks
London, UK

John Hindle
Nashville, TN, USA

Matt Stanton
Swalcliffe, UK

John Smith
Innellan, UK

ISBN 978-3-031-46568-0   ISBN 978-3-031-46569-7   (eBook)
https://doi.org/10.1007/978-3-031-46569-7

This Palgrave Macmillan imprint is published by the registered company Springer Nature Switzerland AG
The registered company address is: Gewerbestrasse 11, 6330 Cham, Switzerland

Paper in this product is recyclable.

Acknowledgements

Our research into service automation, intelligent automation, AI and digital transformation has involved thousands of survey respondents and case research interviewees. Our heartfelt thanks for sharing your knowledge, giving your time so generously and steering us towards an ever-deepening understanding of the technologies, how they are received in organisations and the challenges, risks and success factors. Without you, this research would not have been possible. For the present study, we thank the suppliers, major organisations and customers who generously gave of their time and brains in answering our questions and for providing such wonderful case details.

Introduction

Our general research into IT systems at Knowledge Capital Partners dates back over thirty years, while our focus on new technologies—Cloud, RPA, Intelligent Automation, AI and all of what we call the SMAC/BRAIDA digital technologies (see Chapter 1) date roughly from the emergence of cloud computing around 2010, through the introduction of Robotic Process Automation in 2012, and continuing developments in the major trending technologies, whose adoption will accelerate over the next 7–10 years. We think of these as comprising social media, mobile, analytics, cloud, blockchain, robotics, automation of knowledge work, internet of things, digital fabrication and augmented reality, with a watching brief on 5G, Web 3.0, quantum computing and metaverse. Not surprisingly, we see this as provisional and readily updateable. We research, understand and interpret them in the longer-term context of IT-based innovation and practice as applied to organisations everywhere but predominantly in businesses, non-profit organisations and government.

About the KCP Research Base

Knowledge Capital Partners is an expert research, advisory and communications firm providing independent, evidence-based insight into the impact of intelligent automation and digital technologies on industries, markets and organisations. It develops research-based planning, performance and measurement tools to guide successful technology deployment and navigate organisational challenges faced on the journey to digital transformation. It offers commissioned research services, thought leadership reports, strategy and operational advice and communications development for internal and market use.

We are writing, here, primarily for executives and practitioners and will endeavour to keep the academic apparatus and language to a minimum. Our apologies for when we fail. Nevertheless, readers will need some reassurance that the recommendations and action principles we arrive at are based on detailed, objective, rigorous, independent work. The KCP research base for this book draws upon 935 RPA, 160 Cognitive/AI and 86 detailed digital transformation adoption cases, growing to nearly 1200 cases by mid-2023. This has been supported by annual surveys on these topics across the 2015–2023 period. The cases and surveys cover client and vendor organisations from Europe, the USA, Canada, Australia, the UK, Asia Pacific, South America and Africa. As well as carrying out new research for this book over the 2022–2023 period, we also draw upon our prior publications. In particular, we would mention *Service Automation, Robots and The Future of Work* (2016), *Robotic Process Automation and Risk Mitigation: The*

Definitive Guide (2017), *Robotic Process Automation: The Next Phase* (2018) and *Becoming Strategic with Robotic Process Automation* (2020). There are also multiple work-in-progress papers available on our website, www.Kno wledgeCapitalPartners.com. This includes research not published elsewhere and work prefiguring forthcoming books and publications. For those interested in our research methods, full details can be found in Willcocks, L., Lacity, M. and Gozman, D. (2021), 'Influencing Information Systems practice: The action principles approach applied to robotic process and cognitive automation,' *Journal Of Information Technology*, 36, 3.

Contents

About the Authors

Leslie P. Willcocks is Professor Emeritus at the London School of Economics and Political Science, and Associate Fellow at Green Templeton College Oxford. He has an international reputation for his work on automation; the future of work and skills; global business management and sourcing; digital business; digital transformation; adaptive business strategy and organisational change. He worked in consulting and IT project management for ten years, then in business education at Oxford, Warwick, and the LSE and is now research director at advisory group Knowledge Capital Partners, and Editor-in-Chief of the *Journal of Information Technology*, one of the premier-ranked journals in the field, focusing on technological innovation.

Leslie is co-author of 73 books and has published over 230 refereed papers in journals such as *Harvard Business Review, Sloan Management Review, California Management Review, MIS Quarterly, MISQ Executive* and *Journal of Management Studies*. His work also appears in major media outlets such as *Forbes* magazine and *HBR* online. In 2023, he was in the world's top three most cited researchers in his field. Leslie has delivered company executive programmes worldwide, is a regular keynote speaker at international practitioner and academic conferences, and for 25 years, has been retained as an adviser and expert witness by major corporations and government institutions.

John Hindle is managing partner of Knowledge Capital Partners. He has an extensive international business background as a senior marketing executive and adviser to companies in the US and Europe. He is Vice Chair of the IEEE P2755 Intelligent Process Automation Working Group, a multilateral standards initiative for the growing Intelligent Process Automation industry. John holds a doctoral degree from Vanderbilt University and has held Adjunct Professorships in Human and Organisational Development with Vanderbilt, and International Marketing with New York University in London. He is a past Trustee of Vanderbilt University. John has published many papers on outsourcing, reengineering and automation and is co-author of *Becoming Strategic with Robotic Automation* (SB Publishing, 2019).

Matt Stanton is a marketing professional with over 20 years' of experience across fast-moving consumer goods (FMCG), pharmaceutical and biotechnology sectors. His professional experience spans global marketing roles, providing marketing consultancy services for leading international organisations, along with business start-ups in FMCG, pharmaceutical and biotechnology space. His expertise covers conceptualising marketing strategy, helping organisations find innovative customer revenue streams and growth opportunities and 'stress-testing' business models based on the assessment of internal capabilities and external competitive environment using wargaming methodology. Matt holds an International M.B.A. from NIMBAS, Graduate School of Management, Utrecht, The Netherlands. Matt is a partner of Knowledge Capital Partners.

John Smith is Knowledge Capital Partners' communications specialist. He is an advisor and writer with exceptional experience of working with business leaders on communications critical to the outcome of major changes designed to increase the growth, sustainability and value of their organisations. He offers great insight into the challenges facing organisations that must enable employees, customers, investors and the wider community to trust and support their aims. He has led the process of defining the messages and managing communications channels to stakeholders on enterprise-wide transformations, acquisitions, post-merger integration and the deployment of game-changing technologies. He has worked in this role as an independent consultant with both corporate and public services clients, with emerging and mature businesses, across most major markets, internationally.

List of Figures

1

Where Are We Now? Where Are We Heading?: From RPA to Digital Transformation

The secret to growth in this new era of disruptive technologies is being willing to learn and relearn even if what you knew previously brought you success.

Nicky Verd

Robotic process automation takes the robot out of the human. Intelligent automation and AI try to put the human into the robot. Digital transformation attempts a whole organization change founded on emerging digital technologies. The difficulties rise exponentially across these endeavours.

Leslie Willcocks

Introduction

Our purpose in writing this book is to provide a realistic and reliable guide to planning and deploying successfully the digital technologies that will improve the performance of businesses. Selecting the technology turns out to be the (relatively) easy part. Putting it to work and gaining full value from it is anything but.

To offer such a guide in a market characterised by contested claims, false starts overstated expectations, and underestimated difficulties seems to us to be a useful and timely activity. We bring to the project expert insight into the ways in which transformative technologies gain traction in the world, and work from a strictly evidence-based perspective.

This introduction provides an overview of our topics and concerns, but in a relatively novel way. Over the last year we have conducted many interviews

L. P. Willcocks et al., *Maximizing Value with Automation and Digital Transformation*, https://doi.org/10.1007/978-3-031-46569-7_1

with journalists, magazines, online think tanks, and academic journal editors, who have asked for summaries of research, perspectives on emerging issues, and predictions of how things are likely to turn out over the year, and the next five to ten years. Generally the major focus has been automation and digital technologies, with relative neglect of issues like management and organisations, except when the subjects of job loss and skills shortages arise. Business value also seems to escape attention, partly because, as we often find out, few people, including businesses themselves, actually monitor this carefully.

Here we provide a composite of the questions asked, that takes into account the full range of questions, and not those just asked most frequently. The result is a strong overview of our on-going findings, together with a provocative but valid, and hopefully easy-to-read introduction to the later chapters of this book.

Misunderstandings

Most people know the headlines. But are they right? What are the top misunderstandings about automation and digital transformation circulating in the media?

There are all too many! But we will limit ourselves to three:

The first is the hyperbole about artificial intelligence. 'AI' is such a useful shorthand is it not? But it's also very misleading. Somebody observed that if it's intelligent it's not artificial, and if it's artificial it's not intelligent. That's correct, but you could also say that it's not even artificially intelligent. Much depends on how you abuse the word 'intelligent'! The area is pervaded by the seductive metaphor that computers are like brains and brains operate like computers. And, of course, technology companies and the media ramp up the rhetoric to suggest that there is a lot more in the technology than there really is, or likely to be any time in the next 15 years. At base what we have is machine learning, algorithms, natural language processing, image processing backed by traditional statistics and, really, the two key developments—impressive and growing computing power and memory. This can produce hyper-speed and impressive results for very limited applications. But there is no general-purpose intelligence. It is 'weak, weak AI'. Of 18 sets of skills used at work several studies including our own found only eight fully automatable. Humans have eight distinctive capabilities and composite skills (the automatability of three further sets depend on context and use), and these human skills are increasingly valuable because they are unlikely to be replicable in the next 15 years, if at all.

Secondly, having experienced and worked with information, communication, and now automation and digital technologies since the 1980s, we at KCP are still surprised at how people believe that a tsunami of automation will slip easily, seamlessly and at great speed into our work organisations—for good or ill. That is not at all how it seems to work. Generally speaking, when technology hits an organisation, strange things happen. The technology is rarely seamless. Even so, 25 percent of the challenges are technical, in our experience, and 75 percent are organisational and managerial. The easiest way we have found to communicate this is to talk of the eight-siloed organisation. The silos that inhibit the free flow of data, information and knowledge, and application of technology are: culture; process; legacy technology; data; strategy; skills; structure; and the big one—managerial mind-sets. A major reason these silos exist is where organisations are structured in business divisions and functions that have become self-contained over time. Most organisations are, even today, struggling with going digital. If you ask them how they score on these silos from 1 to 10 (with 10 being very siloed), most, even today, will have a significant number against each of these eight areas. And there you have some key reasons why automation and digital transformation are experienced by so many as so challenging.

Interestingly overlooked, but a real trip wire for going digital is data. Actually, we find 80 percent of organisational data is usually semi-structured or unstructured and not that usable. Usable data for automation technologies may be as little as 15–20 percent. Then you hear about the wonders of big data; it has been nicely said that the dirty secret of big data is that nearly all business data is dirty. For example, it comes preloaded with biases, it's frequently not in a form that is usable, or that you can compare with other data. Given the statistical basis of many algorithms that depend on such things, getting a random sample is much easier stated as a principle than delivered on in practice. All in all, the point is that the data challenge has to be faced before the technological and organisational ones, and the data challenge is far from trivial for most working businesses, let alone something like a major government department like tax or social security, or, in the UK, the NHS (National Health Service).

Even assuming that the organisation has the capabilities to manage the technology into the organisation, you can see that these silos create a very big set of challenges to effective deployment.

Allow us one more. The third misunderstanding relates to the idea that technology is no longer a specialism, needing specialist knowledge and experience. In practice people have been discounting the Information Technology (IT) department since the 1980s. A Sloan Management Review paper in

1985 basically said we were all going to become our own technologists, that the technology would become simpler to operate, and our knowledge would be much greater as well. Farewell to the IT function. That has not happened. The systems are more connected, more invisible, more complex than ever before, and we have become much more reliant on technological experts. When deploying automation technologies, very little can be done at scale, strategically, unless it fits very meticulously with the technology platform of the organisation—its governance structure, protocols, security processes, and technical architecture. The risks of not doing this are now so much greater. The era of the 'citizen technologist' has not arrived. Pointing at the mobile phone as an exemplar of modern user-operated technology is simplistic because it ignores the enormous amount of technology that has to be in place to make that work so seamlessly.

Impact of Automation on the Global Workforce

What is the evidence saying about the impact of automation and digital technologies on the labour force?
   The evidence is surprising. Back in 2014, influenced by a lot of the books and reports coming out then, we first assumed the equation was automation equals job loss. Across our 750 plus cases and in our surveys, we found certainly around 10–13 percent looking for labour displacement. But, looking at the 2015–2023 period, the majority were automating because they were experiencing year-on-year massive increases and intensification in work, they had skills shortages, productivity problems, and/or they could not meet work processing targets.
   Another surprise was that in most cases people did not fear but embraced automation. It helped their work; it took away the painful, deskilled parts; jobs were restructured to include more valuable tasks—for example moving people into dealing directly with customer service problems. All that improves employee morale, and job satisfaction. We once interviewed a CEO in the USA and he summed it up as—the problem with US industry is too many idiot jobs and not enough idiots to fill them. Well automation—RPA and cognitive automation—provide some answers to that one!
   Now we are not saying that all workforces will have these experiences. Some sectors and occupations are going to be hard hit by automation over the next ten years. In particular we expect to see 68 percent of data processing activities automated in the next ten years, 64 percent of data collection, and quite a lot of physical work. But the studies show there will be large gains

for humans in areas such as applying experience and specialist knowledge, interfacing with stakeholders, and managing people. We give other examples below.

It is worth looking at the issue of job loss, which is very misunderstood. Actually, the net job loss from now to 2030—about six major reports including our own agree—is about 1 percent. We have read early reports that quote figures like 47 percent of the US workforce jobs are automatable, then jump to the conclusion that this is what is going to happen. It is not. Leslie Willcocks provided an in-depth analysis of this in a June 2020 paper in the Journal of Information Technology called *Robo-Apocalypse Cancelled? Reframing the Automation and Future of Work Debate*, that gave eight qualifiers to these big estimates. It's not whole jobs but part of jobs that will be automated. These early studies leave out job creation as a result of automation. More recently, McKinsey, for example, found that 18 percent jobs might be lost from automation by 2030 but 17 percent will be gained. There is an over-belief in how easy it is to implement these technologies. Ease is certainly not what we are finding in our studies of the banking, insurance, telecoms, healthcare and energy sectors—in fact, reviewing the studies, it takes anything between 8 and 26 years for a technology to be 90 percent adopted in the developed economies.

There is also a massive over-optimism about how perfectible these technologies can be. Watch the driverless vehicles and ChatGPT spaces for learning experiences on this one. Then there are the macroeconomic factors like ageing populations, productivity shortfalls, skills shortages, that help explain why countries like China and Japan are automating so fast—it is as a coping mechanism in the face of labour shortages, and not to eradicate the labour force. Finally, in our estimates the amount of work to be done globally is not static, as virtually every study assumes, but is in fact increasing at around 8–12 percent per annum annually. Ironically some of this is due to the exponential data explosion, and the problems automation and becoming digital create—the cybersecurity industry is growing exponentially for a reason! Once again automation becomes a coping mechanism, in this case in the face of growing workloads, as we have found across our main 1200 plus case database.

So generally, looking across our research and all the studies, one can create a picture for the next seven years that suggests, admittedly depending on a range of variable factors, that there will be job creation—around 10 percent of jobs in 2030 do not exist in 2023. Some job occupations will shrink so that around 15 percent of the global workforce will change occupations. Around

8–10 percent of jobs may will cease to exist. McKinsey—rightly we think—see 60 percent of jobs will be more than 30 percent automated by 2030, but only around 9 percent of jobs today are wholly automatable. When you think about it, this is not that surprising as most jobs have seen the steady take-up of automation and digital technologies over many years, and this would seem to be a continuation of that trend. Despite the predictions of massive job loss, more recent studies are in line with that of McKinsey mentioned above suggesting that by 2030 there will be a dual impact, with 18 percent jobs lost and 17 percent gained from automation. Our own work suggests that the big issue facing the major economies is not job loss and unemployment, but skills shifts.

A Closer Look: Automation Jobs and Skills

Does automation change jobs and the skills required?

Well, each organisation, and indeed sector, will be different, but the basic trend over the next 10–12 years is a move away from low skills—physical, repetitive, non-technical, non-cognitive, basic human skills—towards digital, technical (STEM) cognitive, distinctively human, medium/high skills. Think in terms of human skills like empathy, teambuilding, leadership, motivation, critical thinking and imagination. We once did this exercise in a hi-tech company and we spent some 40 minutes listing out the skills which humans bring to the workplace that singly—and especially in combination—cannot be inexpensively and easily, if at all, replicated by machines.

Let's not get this wrong. There will still be impacts on the low skills mentioned above. But the rhetoric tends to run away with itself on this. For example look at what is predicted for remote working—will we all turn to home working as a result of the COVID-19 experience? Well in actual fact 60 percent of the US workforce cannot work from home due to the type of work they are doing. By 2023, many organisations were settling around the notion of employees being allowed a maximum of two days homeworking in a five-day week.

The truth is the real issue is not dramatic net job loss, but of massive skills shifts. The world has 95 million surplus low skilled workers, but a 90 million shortage in medium/high skills. As one example, China will probably have to reskill some 220 million workers over the next seven years. On several estimates low skilled workers will go from 44 percent to about 32 percent of the global workforce over the next 10 years. You can see that with this massive transition, large skills shortages will be experienced, which interestingly enough, may well promote further automation.

These major skills gaps will widen without government, corporate and individual interventions. It is pretty clear also that the inequality divides arising from automation/digitalisation will require labour market institutional changes, but as yet it is not clear whether the COVID-19 experience has whetted or actually inhibited the appetite of governments for further interventions into labour markets.

Robotic Process Automation

"What is Robotic Process Automation (RPA), and what is its future role?"

RPA is a fairly straightforward piece of software that replaces a human in fairly simple, repetitive, information processing tasks. In a phrase one of us invented in 2015, *"RPA takes the robot out of the human."* Modern organisations are absolutely full of such 'robot' tasks, so RPA has wide application. RPA software contains pre-configured rules which processes structured input data to produce correct deterministic outcomes. If it cannot produce an answer, it throws out an exception to be handled by a human. Essentially, it's an information processing execution engine. Unlike more advanced tools, it does not 'learn'; it does not contain complex algorithms; it cannot deal with unstructured data, images, or natural language. However, it can work well in a complementary way with both advanced cognitive tools—and also humans—as part of a broader work processing system.

The RPA role historically has been to improve labour productivity by providing a quick, cheap solution that avoids the IT prioritisation queue and delays. RPA does not need specialist developers, it can be configured quickly, and business operations can run it within IT security, protocol and guidelines. Organisations have had a lot of problems moving this kind of use to the next stage, because they tend to treat RPA as a low-level tool, despite the multiple benefits we have been seeing from its application. We will address that below.

But the future that we wrote about in our book, *Becoming Strategic with Robotic Process Automation*, is already being progressed, by, we would say, 20 percent of organisations across sectors. And what we are seeing, firstly, is much greater scaling—trying to apply RPA across the enterprise, and also to end-to-end processes. Some RPA products lend themselves to this more than others, because they reduce the technology integration, security and protocol problems that can arise.

Secondly, there is a shift to using RPA together with cognitive automation tools—for example one cognitive tool might structure the data for use in RPA, and another might take the data from RPA and carry out predictive analytics with that data. The result is that this sort of ecosystem of automation technologies is often called 'intelligent automation'. Some use intelligent automation to mean using RPA and cognitive automation tools in combination. This seems as good a way to go on definitions as any. We baulk at the word 'intelligent', but it is certainly a more intelligent use of automation! Anyway, one result is the creation of an enabling automation platform that can give management a lot more choices, opportunities and faster innovation—not just for internal operations, but also for external relationships with customers, for new services and products, and greater competitiveness. Our research shows that organisations are accelerating their moves in this direction, not least because there is a huge amount of value being left on the table—a further US$1 trillion annually in global banking alone, according to McKinsey. But too many organisations still have a tactical view of these automation technologies, while a lot still find the challenges difficult to surmount—more below on that.

Thirdly, our research throws up leader organisations that see RPA and cognitive automation as part of much larger digital transformation efforts, moving much more work to the cloud, and combining automation technologies with other advanced digital technologies—we call them SMAC/BRAIDA—see Chapter 16—so you will observe increasing links in the future with social media; mobile; analytics; cloud services; blockchain; robotics; automation of knowledge work; Internet of Things, digital fabrication; and augmented reality technologies. Should we add 'metaverse' to that list? Probably not just yet!

RPA—Good for Stakeholders?

"What impacts can RPA have on businesses, and other enterprises?"

At the start of our research, in 2014, there were few RPA users, and we chose to look at only the most successful. What surprised us was the multiple benefits they were getting—many unplanned for, and completely unexpected. Organisations tended to be going for a return on investment, and were looking to lower the cost of operations. They were getting ROI of anything between 30 and 200 percent in the first 2 years, depending on the process. They were saving large amounts of time—what became called the 'hours back to the business' metric. Work was being done faster, cheaper, more accurately, 24 × 7, with strong audit trails.

Then regulated industries like energy, insurance and banking found they could get quicker regulatory compliance when regulators changed the rules, thus avoiding large fines for non-compliance. One UK insurance company, for example, turned to automation after paying out a £8.4 million fine for non-compliance.

Fundamentally, organisations were finding that more work could be done, with the same or less numbers of people. At the same time the customer service and experiences were being addressed and enhanced—queries, insurance policies processed, repayments processed more quickly, more customised treatment. Then employees were appreciating the improvements in their work loads and type of work they were doing. Some were being moved to working directly in the CofE (Centre of Excellence) and becoming specialists in automation.

But as more companies moved into RPA, and our research base widened, we found that this triple win for shareholders, customers and employees was by no means the common experience. Some of the reasons for this were to do with the limitations of the software—not everything called RPA is the same, and some vendor products are difficult to scale to enterprise level. But our second book, '*Robotic Process Automation and Risk Mitigation*', detailed some 41 material risks that emerged, and the many management actions that could be taken to get the triple win. We learned these from the successful companies—we now have some 39 of what we call 'Action Principles', which we have published widely. A lot of managers are still working through how to deal with those risks but our latest work suggests that at the back of it they need a much more strategic vision to get the payoffs that we know are there from applying these automation technologies.

Establishing Digital Leadership

"Which businesses are more likely to gain a lead with automation and digital?"

We have been studying this for the last 9 years, and a clear picture has emerged. The top attribute is that they have an executive team that believes that digital technologies can be transformational and of strategic importance to its business as whole. Actually, the evidence is that these technologies, managed well, do have these characteristics. The more digitised businesses are, the better the financial results they get in their sectors. In fact, getting ahead and being a leader in these technologies is likely to give the organisation a competitive advantage that can become irreversible over a 5–7-year period.

Such an executive team will fund and resource the technology as a long-term strategic investment to build a digital business platform that delivers necessary internal changes much faster, can support new product/service and can seize business opportunities as they arise.

In our book '*Becoming Strategic with Robotic Process Automation*' we detailed all the strong attributes of a digital leader. Getting there, we suggest, involves:

1. Think, then act strategically—the business imperatives dictate your technology investment.
2. Start right—pilot but build always with the end technology platform, architecture and infrastructure in mind. Build the organisational and technical capabilities for delivering and deploying the technology.
3. Institutionalise fast—we had transformation managers who refused to start until they had all the stakeholders on board and the governance structures in place.
4. Remember that if it's strategic, it involves major organisational change, and change management really is the key to the door here, and not just an add-on. At the heart of it is a major cultural change that will take several years to happen.
5. Innovate continuously. For some examples, have a look at Amazon; DBS Bank Singapore; Schneider Electric; Google; TenCent; LexisNexis. Their digital platforms allow for faster, easier and more extensive innovation both internally and in external relationships and business positioning.

Beyond RPA—Intelligent Automation and 'AI'

"What Is intelligent automation, in the scheme of things?"
Over time, especially as more advanced technologies emerge, it became obvious that stand-alone RPA was essential, but offered unnecessary limitations. Intelligent automation consists of more powerful software suites that combine RPA with more advanced technologies—typically business analytics, OCR, intelligent character recognition, natural language processing algorithm and machine learning based software—that can automate or augment tasks that do not have clearly defined rules. We have also called such technology 'cognitive automation' but, as we mentioned above, really do not like

to call such software 'Artificial Intelligence' because we believe the AI label aggrandises what these tools do. With these technologies, inference-based algorithms process data to produce probabilistic outcomes. A variety of tools are in the realm of intelligent automation, such as tools that analyse data based on supervised machine learning, unsupervised machine learning, and deep learning algorithms. While some of the algorithms have been around for decades, only recent advances have provided the computational power needed to execute them on big data. The input data is often unstructured—such as free form text, either written or spoken. For example, at Deakin University, Melbourne we found intelligent automation being used to answer natural language student inquiries. The input data can also be highly structured, such as the pixels in an image. Google's Machine Learning Kit, IPsoft's Amelia, IBM's Watson suite and Expert Systems' Cogito have been older examples of IA tools.

People refer to 'strong' AI as trying to use computers to do what human minds can do. The vast majority of organisations are a long way off that! However, 'AI' is widely and misleadingly used, especially by vendors, as an umbrella term for RPA and cognitive automation (see Fig. 16.2), as well as much more advanced software that has not even made it out of the laboratory. The term 'AI' is now used so ubiquitously that we may have lost the battle on the terminology. That's OK as long as everyone keeps in mind that what we have today, in businesses is 'weak, weak AI'—algorithms and machine learning, driven by massive computing power. This is not 'intelligence', and much of it is in fact 'statistics on steroids'.

So, to summarise, where were organisations with this in 2023? We are seeing organisations increasingly combine RPA with what they call 'AI'— computer systems that seek to simulate and outperform human intelligence, for example analysis, logic, memory, processing—machine learning using structured semi- and unstructured data, computer vision and natural language processing tools, and process discovery and mining, whereby data analysis is used in order to detail and improve business processes. The potential applications are vast, and exciting—witness for example the immense interest in Chat GPT from early 2023—but therein lie also at least three problems: The first is choice. Which processes and applications do we go with? The second is management. How do we productively build the capability to develop, manage and deploy these technologies? The third is how do we control the power and possible downsides, and unanticipated consequences of what we create? It is worth signalling a fourth issue, which we touched on in an earlier question, because it is proving a real stumbling block in a number of large organisations we have been recently researching.

The challenge is data. As much as 80 percent of an organisation's data is 'dark', meaning that the data is un-locatable, untapped, or untagged. Enterprises first have to create new data and clean up dirty data that is missing, duplicate, incorrect, inconsistent or outdated. Enterprise adopters of IA tools also struggle with 'difficult data', which we define as accurate and valid data that is hard for a machine to read, like a fuzzy image, unexpected data types or sophisticated natural language text.

Now intelligent automation/'AI' applications generally require large data training sets, and thereafter are set up to deal with massive amounts of variable data. Processing power and memory race to keep up. If a great deal of data is not fit for purpose, then this bad data will create misleading algorithms and results. The idea that very big samples solve the problem—what is called 'Big Data'—as used in ChatGPT, for example, is quite a naïve view of the statistics involved. It is not really possible to correct for bad data. And, as we said before, the dirty secret of Big Data is that most data *is* dirty.

Digital Transformation

"Digital transformation is one of those terms where so many people have said it for so long. It's a buzz word in danger of losing meaning. What does digital transformation really mean?"

Well, it means what most people AREN'T doing! But they use the phrase to describe what they ARE doing. So, digital transformation is a whole organisation, radical restructuring of people, processes data and technology, to become a digital business, essentially. Most organisations are really having problems with that, because, as we mentioned before, they are heavily siloed. Those are the legacy silos of the organisation because the organisation was set up to run in a different way, and then you bring in technologies which allow you to run in a different way again. But you need to reorganise and restructure in order to allow those technologies to reveal their potential. So that's what digital transformation is to us—breaking those silos down and utilising the potential of digital technology to become digitally propelled businesses.

But we think a lot of people have redefined it. So, for example in the downturn in 2023, all the signs suggested that organisations were really cutting down on technology investments, including the hi-tech businesses which had over-expanded, thinking that the technology would take off. The COVID-19 experience gave them positive, warm signals about the role of technology in a future economy. By early 2023 we found businesses talking more about *cost* transformation through digital technologies. And that's the way that digital transformation was being reconfigured for most organisations during 2023. The terminology has been the same, but the purpose morphed into: how can we use digital transformations to keep our costs under control and achieve economies and process improvement without really doing those fantastic strategic things or using the top leading-edge technologies that people have been trying to sell us for the previous decade.

So that's what where we were by mid-2023, Even the go-head companies, the leaders, seemed to be coming back into defining digital transformation in practice as cost transformation. So, they were seeing a role for RPA, and intelligent automation, and AI, but it has been much more redirected into narrow goals.

In early 2023 we reviewed how representative organisations in our KCP database were dealing with current economic conditions. We found that organisations had responded to post-COVID-19 conditions in four ways. By 2023, some 20 percent were 'Sweating the Assets'—that is, making the most of existing technologies to achieve short-term business survival goals, including customer retention, cash flow maintenance, and cost cutting. The figure for July 2022 had been 35 percent. Meanwhile, another 45 percent (30 percent in July 2022) were in a more advantageous position, pursuing the short-term goal of 'Underpinning Today's Business' towards which new automation and digital technology investments were directed.

A further 20 percent were 'Slowing the Digital Strategy'—they had a long-term digital strategy in place, but it was proceeding at a slower rate, and some of the new digital investments were being channelled into short-term objectives. Only 15 percent of organisations were 'Adapting Strategy and Building Resilience'. These had a long-term, adaptive strategy and were future-proofing and building resilience with large new digital investments.

What can we learn from this, going forward? Clearly one can see business imperatives driving technology investments—no bad thing. But short-termism in the digital arena can be dangerous in eroding resilience and sustainability the next time economic conditions are adverse. Moreover, we think it is leading to a digital divide in each sector, creating competitive advantage for the long-term digital investors that could become irreversible. Digital leaders are emerging representing 15–24 percent of organisations, depending on sector. Several studies combined with our own work show the general picture that they gain something like 20 percent profit gains from their digital investments, representing twice more than digital laggards. They also get 20–25 percent more revenue, higher market valuations than their peers, and offer superior dividends.

A general finding is that this performance gap between the best and worst digital performers was widening between 2018 and 2023. It's not just that digital leaders are investing more strategically and on a greater scale. They are also improving faster than their competitors in their ability to do digital. Thus studies, including our own, show that they test ideas more quickly than before, move faster on digital than before, are scaling over five-year periods faster than before, whilst improving their digital execution capabilities along the way. This trend will continue, we suggest, unless organisations seriously adopt the management guidelines and key digital capabilities we detail throughout this book.

An Automation Sweet Spot

"Looking ahead, where is the sweet spot that can make automation and digital transformation work?"

This needs a complicated answer, rather than a clear one, but we think it's about where you start from. We recall, back in the 1990s, there was a big thing about business process reengineering, and radical reform. It was overstated in terms of the radical 'digging up the roots' change required, but we do think that you have to start with the processes of the organisation—identify what the core processes are, and how you're going to digitise them, and then the other processes as well.

So, for us, it begins with process discovery, really. What are the core processes given the sort of business we want, and how do we utilise digital technologies to leverage those processes? And how do we organise our data to run those processes? That, to us, is the fundamental starting point. And how do you do that in an organisation? We don't think you can do it from a bottom-up approach, but we don't think you can do it from a top-down approach—which is where radical, business process reengineering went wrong. A 75 percent failure rate tells you something about radical re-engineering, but recall also that digital transformation records a 2023 'failure' rate of around 65 percent. We think you have to do it from the middle. Because somehow, it's a shorter communication path, and it links with the bottom-up people who are saying, *"Well, we're really having problems here,"* and the top-down, people saying, *"Well, strategically, we need to go here,"* so you then say, *"Well, what processes do we need in place?"* and you invent some form of governance that allows you to do process discovery in different parts of the business. What are key business processes to deliver on your business imperatives? And then you start organising your data and your technology around that.

Robotic process automation has been—and in some places still is being—sold as a bottom-up approach, and all too frequently it's a one-off process. You know: *"We've automated this and look at the benefits we're getting from this process. So, let's look at another process!"* But do we ever look at it from a more holistic perspective? People tend to automate the processes easiest to automate first, but is it more strategic to look at the ones that are going to give you the most value?

Centres of Excellence

"So, do you think Centres of Excellence is the way to go?"

Yes. We concluded this between 2016 and 2019, and then we saw the Centres of Excellence moving from just RPA Centres of Excellence, to automation Centres of Excellence, and then, certainly in the digital leaders, linking up with digital transformation efforts. More recently, what we also found in many big companies was that there was this hole, where automation initiatives were not connecting up with the team running the digital transformation initiatives. New siloes were occurring in these big organisations, so the digital transformation was tending to be driven top-down, and the automation tended to be driven from the Centre of Excellence at best, and very often bottom-up (see Chapter 17 on the Blind Spot). And there was little integration of the automation technologies—though people were increasingly getting into intelligent automation.

So, during 2023 we were finding that people were going digital but in a relative fragmented way, and that's a problem. In terms of the sweet spots, first they have to get fundamental about process, which is best driven from the middle of the organisation, linking it with business imperatives. Then you organise your data and technology around that. And then you create the link from the middle of the organisation with the digital transformation efforts in order to start integrating what you're doing with the automation, and what you're doing with the other technologies. This is the sweet spot because digital technologies really gain their massive value from integrating with other digital technologies. When you start linking blockchain with automation; when you start linking seriously advanced automation technologies with RPA, you suddenly start getting much more value. You start getting what we call an automation platform—then eventually a digital platform. And it's the platform that enables, not just new business processes, not just new business development, but new businesses can spin off the digital platform, as we've seen with many cases—not least Amazon.

So, as a summary answer to the question, one of our fundamental findings from looking at RPA from 2014 to 2023 was that you have to centralise governance more than you think. That's why we keep saying you have to start in the middle. You can't really start bottom-up. You know it's nice to try out a few things, but that's a very low-level fruit that you're picking there and you have to start much higher up in the organisation in terms of governance. And on a bigger scale of whole organisation digital transformation, assuming this has to be, in fact an evolution, governance from the middle becomes a key factor, particularly in organising processes, people technology and data, and integrating bottom-up with top-down perspectives.

Conclusion

"What is the future of automation and digital technologies?"

First, here are some figures, then we will provide some comments. There are many estimates of the combined market for RPA, Intelligent Automation and 'AI'. Statista suggest that the market is expected to show strong growth in the coming decade. The 2022 market value of nearly US$100 billion is expected to grow 20-fold by 2030, up to nearly US$2 trillion. On a Research and Markets estimate the global artificial intelligence market size was US$136.55 billion in 2022 and expected to reach US$1,812 billion by 2030, expanding at a CAGR of 37.3 percent from 2023 to 2030.

If these estimates seem high, then be aware that the AI market is being defined as covering a huge number of industries. Everything from supply chains, marketing, product making, research, analysis, and more are fields that will in some aspect adopt artificial intelligence within their business structures. Chatbots, image generating AI, mobile applications, massive increases in processing power and memory are all among the major trends improving AI in the coming years. According to these reports, the advent of big data is expected to be the cause of AI market growth, as a large volume of data is needed to be captured, stored, and analysed. The increasing demand for image processing and identification is also expected to drive industry growth.

Looking more broadly, Grand View Research evaluated the global digital transformation market size at US$731.13 billion in 2022 and anticipated a CAGR of 26.7 percent from 2023 to 2030. They attribute market growth to the growing adoption of cutting-edge technologies such as cloud, big data analytics, and Artificial Intelligence (AI), among others, which has caused in the exponential growth of all size of businesses across the globe.

Our commentary on these stats is a little less starry-eyed. Predicting the future about technologies, even to five years, is a really risky business! That does not mean that it cannot be informative, especially if you put forward possible futures, rather than THE future. It's probably best to look at trends and where they can lead to and what can upset them. Undoubtedly, these reports are right to suggest a continuing exponential data explosion, fuelled by, and also accelerating the development and application of these automation and digital technologies. Major driving forces will be the productivity shortfalls, skills shortages and economic growth needs that will be increasingly experienced across organisations, sectors and economies to various degrees.

However, accepting these as major trends, how is this all going to happen?

Firstly, technology spend does not equate with value. All organisations have been spending on automation and digital technologies. However, just looking at businesses, we are finding that roughly about 90 percent of the value is going to around 20 percent of the organisations, depending on sector, but even those are securing on average around 68 percent of the value available to the organisation. On a global scale, despite the US$2 trillion expenditure by 2030, many trillions of dollars could be left on the table.

Secondly, we consistently find organisational ability and readiness to develop, deploy and drive value from these technologies is assumed, and so widely overestimated. In practice organisational capability for technological change is not unlimited, and in many cases may well have been eroding over the last few years. This is very likely to continue as a trend. Skills shortages will contribute; likewise, the technologies accelerating at a faster pace than organisations can adopt them. Also factor in legacy systems, siloes in organisations, the challenges of change management and moving to digital cultures, and the fact that many organisations may have limited 'absorptive capacity' for digital transformation. All this suggests a slower, more evolutionary path than the digital technology 'tsunami' so often predicted.

Thirdly, we need to take account of other 'drag' factors. Firstly, as data and technologies become more ubiquitous, they also impact more widely—for good or ill. We cannot afford to underestimate the role of misleading data, poor design, bad actors, and unanticipated ethical and social responsibility challenges. With these technologies, 'can' does not automatically equate with should'. Such 'drag' factors may well inhibit adoption, and prompt a further 'drag', namely more audit and regulation—already almost a decade behind where accelerating technologies have taken us.

But fourthly, none of this is a gospel of despair, more one of realism. Organisations *can* learn from digital leaders, who *are* making headway. As we have consistently found over 35 years of working in research, management *does* make a significant difference, but the primary objective has to be not to drive automation, but to drive value *with* automation, and likewise with the other digital technologies coming on stream. As Peter Drucker once said: *"The best way to predict your future is to create it."*

Part I

Robotic Process Automation

2

A Strategic Approach to Robotic Process Automation

All can see the tactics whereby I conquer, but what none can see is the strategy out of which victory is evolved.
Sun Tsu, '*The Art of War*'.

Introduction

In this chapter, we answer the key question: *"What explains the superior outcomes automation leaders are getting against several market trends?"* Our research shows that the single most important factor in achieving superior outcomes—one that shapes and informs all RPA-related activities—is the adoption of a strategic approach to the introduction and management of RPA within the enterprise. Combining all our research streams, we distil out seven attributes of the truly strategic performer in the RPA space—see Fig. 2.1. Six of these are well supported by the evidence. The seventh—focusing on total value of ownership (TVO)—is still under-utilised, and represents a real opportunity to develop strategic behaviour further.

Strategy *vs.* Operational Quick Wins

Leading companies observe a fundamental rule: business strategy drives RPA investments. In the case of RPA this does not necessarily happen immediately. RPA historically has been seen as a tactical, quick-win tool to achieve

L. P. Willcocks et al., *Maximizing Value with Automation and Digital Transformation*, https://doi.org/10.1007/978-3-031-46569-7_2

> 1. Strategy *vs.* Operational Quick Wins
>
> 2. Culturally Imbedded *vs.* IT As Usual
>
> 3. Planning *vs.* Opportunism
>
> 4. Programme Governance *vs.* Project Delivery
>
> 5. Platform *vs.* Tool
>
> 6. Change Management *vs.* Silo Tolerance
>
> 7. Measurement: Using TVO Metrics

Fig. 2.1 Becoming strategic with automation

business benefits and bypass the long IT work queue. Many RPA tools were set up with precisely this aim in mind, and inherit design limitations when clients attempt to scale them to achieve bigger business goals. Moving from a tactical focus on costs to multi-faceted strategic impacts often follows a typical pattern (see Fig. 2.2). Many RPA users move, sometimes painfully, through Phases 1 and 2, to get to Phase 3. Value increases from initial limited efficiency objectives, through to effectiveness and enablement goals and payoffs. Pioneers we studied over the years—like Telefonica O2, RWE npower, innogy SE Business Solutions, Barclays Bank, and Shop Direct—matured their own strategic understanding over time and came to operate with Phase 3/4 mind-sets.

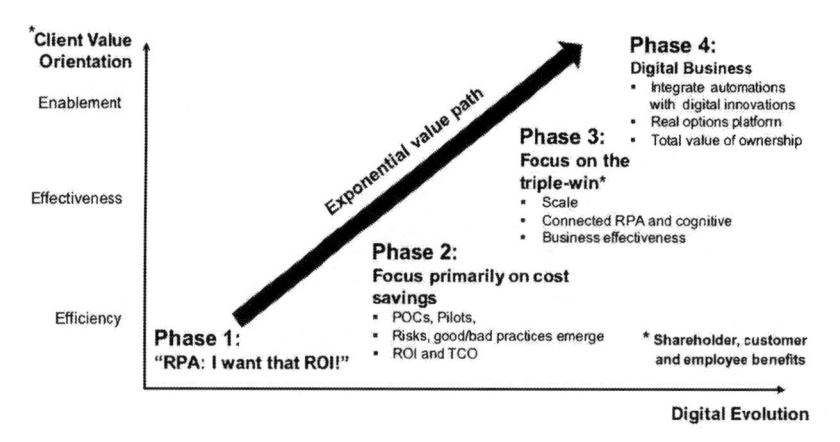

Fig. 2.2 The automation path to value

Building on client experiences, it is possible for companies to accelerate their learning and kick-start at Phase 3. A very early example was USA-based bank BNY Mellon, beginning their journey in early 2016. By mid–2017 they had over 200 robots in production and had automated more than 100 processes.

Where does business strategy come in? IT investments are always best driven by business imperatives. RPA is no different. We found Phase 3 clients going for a triple win of shareholder, customer and employee value. The secret here was the higher aspiration. Clients aimed for and were getting multiple business benefits but were producing also unexpected returns, for example discovering much better regulatory compliance, products quicker to market, enhanced customer journeys and increased employee skills and recognition. By 2023 automation leaders had even higher aspirations, seeking to deploy RPA across the enterprise, linking with other intelligent automation and 'AI'-based applications, and with other emerging SMAC/BRAIDA technologies.

Culturally Imbedded *vs.* 'IT as Usual'

The longstanding finding on executive support for IT investments generally is reinforced by our RPA research—automation as transformative must have cultural adoption by the C-suite. It cannot be merely left to the IT function. This manifests itself in senior executive behaviour. They sponsor service automation and act as project champions. They see RPA as a strategic business project and provide the requisite financial and human resources. They communicate clearly on automation, and ensure that governance and project structure are in place. They protect developments when they run into difficulties. A prime example from our early client studies was Xchanging, where, in 2014, CEO Ken Lever promoted 'putting technology at our core' as an annual report message. By June 2015, Xchanging had automated 14 core processes with a range of significant business benefits. (For more on digital culture see Chapter 21.)

Recent data from clients extends this picture into adoption practices. Over 80 percent of RPA users drive automation from a centralised Centre of Excellence, or top-down through a senior executive responsible for multiple business units. Clients point out that it is difficult to scale and gain the really significant *strategic* benefits from RPA without top-down management and senior executive support. The clients we found deploying RPA locally or through the IT department were at earlier stages of their automation journeys, and not deploying the full capabilities of the RPA platform.

Planning *vs.* Opportunism

Jon Theuerkauf, at the time managing director and group head of performance excellence at BNY Mellon, stated as one of his principles: *"begin with the end in mind."* More precisely, looking at that case, BNY Mellon planned for the mid-term and long-term end-points, and recognised that the 'end-point' would be continually redefined. We have found this typical of clients with a strategic mind-set. Much depends, of course, on what that end-point is defined as.

During 2017, most defined the end-point as establishing an RPA, then an automation Centre of Excellence focused on applying several technologies such as RPA, cognitive, and analytics. By 2018 we found most clients treating RPA as part of a larger automation or larger digital business strategy. Companies like American Express, IBM, BNY Mellon, ING, Nordea, and Siemens planned to start slow, then scale fast. They looked for a rich business value proposition. Such clients, and subsequent successful ones, aimed for high return on investment (ROI)—our research continues to find evidence of 30–200 percent first year ROIs, depending on process. But they also look explicitly for, and get, 'triple wins', including typically, improved service speed; consistency and quality; faster deployment of new services; cost savings; improved regulatory compliance; more efficient processes; differentiating customer experiences; and more flexible, satisfied workforces. In terms of Phase 3, clients have been increasingly enhancing RPA usage by adopting complementary cognitive technologies. We found early examples at Zurich Insurance, in the claim validation process, and at KPMG, in audit, business generation and risk assurance validation processes. Such companies, typically, also plan carefully across the automation life-cycle—from strategy to maturity—to mitigate the 41 material risks likely to be encountered in any major automation program.

Program Governance *vs.* Project Delivery

A common mistake—even in late 2023—has been to treat RPA as just another piece of software. This leads on to limited governance arrangements—at best adopting standard project management techniques—and seeing scaling as just buying more software to spread across more processes, with little IT engagement. By 2018 many clients, particularly those deploying robotic desktop automation (RDA), found this inhibited both scaling and deploying RPA as a foundation for further service automation and digital transformation.

Our case research shows that leading RPA users across sectors take a different route, and, like Siemens, innogy SE and BNY Mellon for example, have seen RPA as potentially more transformational. The constitution ('rules of the game') for automation is formulated on Day One, and covers decision-making and responsibilities for technology, process, data, business and resources. In fact, with some versions of today's RPA, clients get built into the software, a lot of technical governance covering security, compliance, change management, ease of integration with infrastructure, and ease of fit with enterprise applications. Some vendors and specialist consultancies also set out a detailed robotic operating framework that stipulates many enabling and policing rules. Taking development methodology as an example, one IBM executive said the rules may take a lot of time and effort to follow, but *"the reasons for taking a rigorous approach become obvious when you fail to do it, and run into serious, costly problems that could have been circumvented by up front analysis and design."* In our book, *'Becoming Strategic with Robotic Process Automation'*, we also detail the vital role of the IT department in governance and making RPA function optimally. These governance features help, we think, to explain why there is a relatively mixed response to RPA. Not all RPA software is the same. We found those clients using enterprise RPA—designed to fit across the entries and with existing systems—much more positive about the technical platform's scalability, adaptability, security, ease of learning, and speed to deployment.

Platform *vs.* Tool

The requirement for such governance comes from seeing RPA as a platform, and increasingly as part of a larger digital platform, rather than just another automation 'tool'. The 'tool' view sees RPA being side-lined and overtaken by more advanced cognitive automation tools for image recognition, natural language processing, machine learning, and algorithmic reasoning, driven by huge advances in computing power and storage. Conversely, we found that, in digital leaders, RPA is utilised as part of a continuum of complementary automation and digital technologies supporting digital transformation of the enterprise. Many RPA vendors now suggest that their products provide an 'enterprise platform' with cloud deployment. Where these are truly functional, we found clients citing, in particular, enterprise-wide scalability, low coding requirement, strong security, and design for enterprise integration.

Change Management *vs.* Silo Tolerance

Used opportunistically, RPA tools can gain quick wins, but too often they have been deployed as a 'sticking plaster', or 'Elastoplast', on pain points in the organisation. This has the advantage of not having to deal with change management issues, but the serious disadvantage of turning down the transformation potential of automation, and consequent strategic benefits. But most organisations are surprisingly heavily siloed, not just in terms of structure, but just about everything else. Anecdotally we regularly ask business audiences to benchmark the extent to which their organisations are siloed ('1 not siloed, 10 very siloed') on eight criteria—structure; strategy; culture; data; technology; processes; skills; and managerial mind-sets. Invariably the organisations score 4 or above not just on one, but on nearly all the criteria. Across 2019–2023, as RPA adopters increasingly scaled to reap more benefits, we found them encountering ever bigger challenges on change management.

Amongst digital leaders, we found senior executives recognising early the transformation potential of RPA, and explicitly managing the change implications for data, technology, people, processes, and structures. Siemens provides us with an early example of bringing these and our other points together. For its shared services, Siemens established a global RPA Centre of Excellence in mid-2017, to define a global approach. It looked to integrate RPA with the business process/management/operations platform and enterprise platform globally. Critical success factors included integrating RPA into a broader automation strategy, alignment with process governance, C-level support with risk capital, and process optimisation being combined with RPA. Other critical success factors included partnering with IT and external partners; developing expertise in automation and process optimisation; clear governance and operating model, a centralised framework for IT architecture and infrastructure, and stakeholder communications and change management.

As many clients have found, particularly important is getting early stakeholder buy-in—from business operations managers, IT, employees and senior executives. This, we find, involves fully resourcing change management capability, messaging the purpose and value of RPA to staff, and ensuring strategic alignment, new competencies, and changes are institutionalised and imbedded in work practices. On messaging, as discussed in Chapter 1, the general effect of automation has not been job loss, but skills change and skills augmentation. However, these intentions do need to be recognised by those likely to be affected. The key issue here is communicating clearly, honestly and early what is likely to happen to jobs as this is a key worry issue for employees (see also Chapter 21).

Measurement: ROI *vs.* TCO *vs.* TVO

The evaluation of IT investments has always been problematic. At the same time, getting the right measurement system has been a major key to driving business value. In the past organisations have tended not to fully investigate risk and potential costs, understated knock-on cost of operations and maintenance, and not properly accounted for rising human and organisational costs. Typically, we found that organisations using traditional ROI cost/ benefit analysis understated real costs, which frequently exceeded technical costs by 300–400 percent. Our evidence is that many RPA users are committing the same mistakes. One response has been to take much more seriously something called Total Cost of Ownership—that is estimating the expenses associated with purchasing, deploying, using and retiring a product or piece of software. TCO, or actual cost, quantifies the cost of the purchase across the product's entire lifecycle. This can then be compared to the ROI calculation. But with emerging digital technologies this only gets us so far.

The real limitation so far with RPA, automation and digital technology assessment has been in establishing benefits. Most methods greatly underestimate the potential, through scaling and connectivity of these technologies not just to create efficiencies ('doing things right') and effectiveness ('doing the right thing') but offering enablement (supporting new products services, processes and business models). One 2017 Forester Research study created a composite organisation from two client organisations and estimated quantified benefits of US$49.19 million over three years. However, these were attributable to labour savings and endogenous recruitment, training and facilities savings. The study registered cost savings, and also referenced unquantified benefits in top-line revenue from improved customer satisfaction, and in lower security and compliance costs. But the latter were discounted from the final calculation of benefits. In our view RPA, and indeed automation, measurement needs to go much further.

A later chapter reports on applying a measure we call **Total Value of Ownership (TVO)**. As a placeholder, we reference the framework in Fig. 2.3.

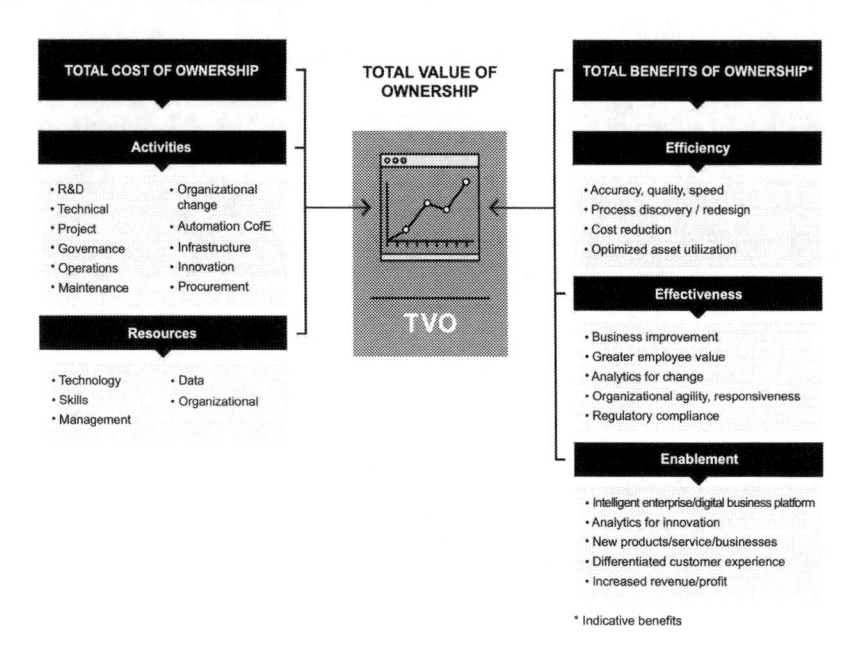

Fig. 2.3 The KCP total value of ownership framework

With this concept, the objective is to ensure that business cases for service automation are driven by (1) total costs, (2) multiple expected business benefits and (3) the strategic returns from future business and technical options made possible by RPA.

Conclusion

This chapter focused on the critical importance of taking a strategic approach to RPA, in the context of developing further 'connected RPA' with more advanced automation and digital technologies. This is the foundation for managing across the RPA life-cycle, as covered in our book, '*Becoming Strategic with Robotic Process Automation*.' This delineates 39 management action principles that cover:

- Resolving key selection challenges: process, vendor, platform, sourcing.
- Gaining stakeholder buy-in and establishing governance.
- Change management and capability development: people, process and technology.
- The path to maturity.

Taken together with our 41 risk mitigation principles (see Chapter 1), these management actions do not just reduce the risk of failure, but enable and ensure maximum enterprise value from RPA as part of a transformational digital business platform.

3

Robotic Process Automation: Just Add Imagination

"Creativity is just connecting things."
Steve Jobs

"Creativity, after all, is a lot like LEGO."
Maria Popova, Smithsonian Institute

Introduction

It can start with a problem … a challenge … an opportunity … an idea … or just curiosity. Whatever the impulse that sparks the imagination, it springs from the human instinct to invent, to challenge, to solve, to create something new. We begin where we are, of course, with the tools we have on our workbench or within our reach. But imagination gives us the freedom to combine them in new ways, to create new capabilities, invent new services, and generate new value. And it's happening all around us, fueled by a profusion of new tools.

As Google's Chief Economist Hal Varian put it: *"Every now and then a technology, or set of technologies, comes along that offers a rich set of components that can be combined and recombined to create new products. The arrival of these components then sets off a technology boom as innovators work through the possibilities."*

© The Author(s), under exclusive license to Springer Nature Switzerland AG 2024
L. P. Willcocks et al., *Maximizing Value with Automation and Digital Transformation*, https://doi.org/10.1007/978-3-031-46569-7_3

In this chapter we focus on robotic process automation (RPA), and counter the perspective that treats RPA as a tactical, often stand-alone tool, which misses out on the potential immense value from combinatory innovation.

Combinatory Innovation

We've seen this before. In the early 1900s, the invention of the automobile married the internal combustion engine with a combination of wheels and gears originally developed to harness water power. The internal combustion engine itself was a combinatory innovation, marrying Volta's electric pistol with Newcomen's piston and cylinder, and Venturi's jet principle. The automobile changed our experience of time and distance, as well as social and economic organisation, and created massive economic growth. Today, as innovative battery technologies and artificial intelligence-enabled, eco-friendly, driverless cars, this global industry is experiencing another wave of combinatory innovation.

Or take the Internet, born in the 1960s as ARPANET, a US-only academic research and defense network that combined packet switching technology, routers, and a few key messaging and inter-connection protocols. When Tim Berners-Lee developed the HTML and HTTP hypertext protocols in 1993, Mark Andreesen used them to create the first web browser—Mosaic—and the race was on, creating what one observer calls a 'global motherboard' enabling services to be built, hosted and delivered virtually anywhere on the planet.

Both the automobile and the Internet demonstrate what economist Brian Arthur calls the phenomenon of 'increasing returns'—the value of a new technology lies not just in what it does, but in what further technologies and solutions it will lead to. Every new technology becomes a building block for future technologies and services.

Today we're in the early stages of a new era of combinatory innovation, marrying earlier technologies and discoveries with new capabilities. Witness the exponential adoption of robotic process automation, artificial intelligence, cloud computing, deep analytics and a growing array of cognitive platforms—natural language processing, speech recognition, computer vision, sentiment and risk analysis, machine learning and reasoning and more.

What makes this different from previous cycles of combinatory innovation is that the components of this revolution aren't physical devices. They're concepts, standards, protocols, programming languages, and software …

which means no manufacturing lines, no inventory management, no shipping delays. As Varian notes of the Internet, *"you never run out of HTML, just like you never run out of e-mail."*

Leading businesses around the world are combining these technologies to free themselves from the complexities and limitations of legacy systems, with innovative solutions that eliminate drudgery, improve productivity and compliance, and create new services and markets through powerful customer experiences. And the impact can be compelling, as a few examples that use RPA as a foundation show:

* A bank that reduced customer wait time for cheque approvals from 7 minutes to 20 seconds using a combination of Connected RPA and AI vision tools to perform 38 automated checks on every signature with 100 percent accuracy.
* A pharmaceutical company that used RPA, Optical Character Recognition, analytics and data visualisation tools to greatly reduce clinical cycle times, product labelling and shipment reconciliations, rushing much-needed medicines to patients globally.
* A global service provider that integrated Connected RPA with a range of security, monitoring, management, repair, and billing systems to provide a fully automated business continuity service and a secure, flexible, pay-per-minute RPA-as-a-service offering.
* An online retailer that integrated Connected RPA with analytics and dispatching systems to identify and intercept fraudulently ordered goods in real time on delivery vehicles *en-route* to their destinations.
* A bank facing a compliance requirement on an impossible deadline integrated Connected RPA, analytics and customer contact technologies to refresh and cleanse 3.5 million customer records, performing the equivalent workload of 650 FTEs.

A large caveat is in order here, however. The McKinsey Global Institute forecasted in one recent report that the gains from intelligent automation and combinatory innovation will not be evenly distributed: *"The pace of AI adoption and the extent to which companies choose to use AI for innovation, rather than efficiency gains alone, are likely to have a large impact on economic outcomes."* By 2030, McKinsey forecasts, *"front-runners (companies that fully absorb AI tools across their enterprises over the next five to seven years) … could potentially double their cash flow"* over their current run rate. Laggards, on the other hand, may experience as much as a 20 percent decline in their cash flows from today's levels.

The driving force underpinning this new era of combinatory innovation, of course, is data—what one observer has called the 'oil' of the new economy. Data is simultaneously a raw material—the source of potentially unlimited insights—and an accelerant—our economic and social fuel. Data, in other words, is an emergent, expanding, self-propelling and self-renewing resource that's re-shaping every industry.

So how are companies using all this data and all these new capabilities? What patterns of combinatory innovation are emerging? What industries are leading the way? What kinds of problems are they solving? What new services and value are they creating? What are they—and we—learning?

First of all, our research suggests that enterprise-grade RPA is emerging as a strategic technology—providing a flexible integration platform and execution capability for the growing range of cognitive and AI applications across the full range of enterprise processes. The most innovative and powerful examples of combinatory innovation in our research share this common denominator—what we might call Connected RPA.

Secondly, early-adopting industries have tended to be the most data-intensive, with high regulatory, compliance, and customer sensitivities—financial services, utilities, retail, infrastructure and public services—as well as organisations that are intentional and adept at using cloud architectures to enable dynamic access to innovative capabilities and services.

Thirdly, leading organisations are achieving notable business value quite early in their implementations by focusing on where value is migrating and how the customer experience can be improved, where new technology can help them respond to unexpected adverse events (e.g., non-compliance penalties), or where there may be overlooked opportunities and unnoticed leverage points relative to the competition—all stimulating imagination, all leading to innovation.

Finally, the most impressive and powerful examples of combinatory innovation are found in enterprises that deliberately seek to capture what we have called the 'triple win'—for shareholders, customers and employees—by approaching their investment as a transformational opportunity rather than focusing solely on operational efficiencies.

Conclusion

In later chapters we will focus on sectors where taking a combinatory innovation approach has yielded outsized strategic gains and/or enabled performance that could not have been achieved with traditional approaches, beginning with financial services, utilities, retail, and infrastructure and public services.

4

RPA in Financial Services

"Money makes the world go around
  A mark, a yen, a buck, or a pound
  Is all that makes the world go around."
  Cabaret (1966)

"Time is money."
  Benjamin Franklin (1748)

Introduction

In his hugely entertaining best-seller, '*Flash Boys*', Michael Lewis tells the story of Dan Spivey, a stockbroker-turned-entrepreneur who built a straight-line fiber optic link between Chicago and New York City that allowed brokers to gain a few milliseconds advantage over their competitors in executing financial trades (roughly a tenth of the time it takes you to blink your eye).

- The cost to participating brokers? US$14 million for a 5-year access agreement.
- The value? Billions of dollars in transaction revenues and client wealth.
- The key? Reducing transaction latency—the time lag between order and fulfillment.
- The lesson? Time is money.

© The Author(s), under exclusive license to Springer Nature
Switzerland AG 2024
L. P. Willcocks et al., *Maximizing Value with Automation and Digital Transformation*,
https://doi.org/10.1007/978-3-031-46569-7_4

Competitive survival in the financial services industry depends on achieving near-real-time fulfillment, and technology has long been the weapon of choice. The industry was among the first to deploy information systems at scale. But ironically, it's among the most challenged today— saddled with generations of legacy systems and IT stacks, with data trapped behind access firewalls and security protocols. Enter, amongst other technologies, RPA.

RPA Applications

Robotic Process Automation (RPA) has proved a powerful first responder in the latest quest for competitive advantage. Banks, insurance companies, brokerages, and trading houses have used it to, in our phrase, *"take the robot out of the human,"* deploying software robots to transact and move information securely at a speed and scale beyond human achievement. But leading adopters have also realised that RPA could be much more strategic.

With creativity and imagination, they're applying the principle of combinatory innovation to streamline operations, improve customer service and create new value. Under their leadership, RPA is emerging as a strategic enterprise platform, connecting and integrating powerful new technologies—AI, cognitive, and machine learning tools—enabling digital and human workforces to interact, support and complement each other. And it's happening all over the world, as these examples show …

A bank in the Middle East has been using imagination and combinatory innovation to address multiple time-critical, high-risk business requirements, accelerating processing times and delighting customers, employees and shareholders.

- Cheque clearing: To speed up cheque approvals, the bank integrated RPA with powerful AI vision tools. A Connected-RPA digital worker invokes 30 different AI vision algorithms to perform 38 checks on every signature, reducing processing time from 9 minutes to 20 seconds, with 100 percent accuracy.
- Fraud: To meet strict Anti-Money-Laundering (AML) transaction monitoring requirements, the bank integrated RPA with analytics and customer contact tools. A Connected-RPA digital worker receives suspicious transaction alerts from an analytics tool, creates and populates a case file, extracts data from multiple bank records, contacts the customer for information, identifies the suspected fraud and presents the case to a human worker for

decision. The solution is delivering a 40 percent improvement in efficiency and 30 percent improvement in accuracy over manual processes.

- Police security checks: To improve response time to police requests for background checks and financial clearance letters, the bank designed a new process integrating RPA, Business Process Management (BPM), cloud translation and customer contact tools. Customers now upload their requests and documents via the police website. A Connected-RPA digital worker pulls the request, collects customer documents and bank records, uploads the request to a cloud service for Arabic-to-English translation. After review by a bank official, the digital worker generates a response letter in English, translates it to Arabic and sends it to the police. The customer receives updates throughout via text. This combinatory innovation cuts customer waiting time from 96 to 2 hours, with 100 percent accuracy.

A US commercial real estate loan-sizing service provider is applying imagination to improve response times and accuracy for customers and to enable its staff to serve clients as true consultants. The new process has enabled the firm to build a distinctive competitive edge.

- Using connected-RPA, the firm integrated BPM, computer vision, machine learning and customer contact tools to meet 8,000 loan sizing requests annually, involving more than 14,000 documents with 24-hour turnaround times. A connected-RPA digital worker receives requests from the BPM platform, logs into client and third-party systems to retrieve evaluation templates and builds comprehensive work files, collecting and storing relevant documents and data on each sizing request.
- Using a custom computer vision tool, the digital worker then extracts data from unstructured documents (financial statements, rent rolls, etc.), validates all tables and texts, completes and uploads the template, and e-mails the client to confirm.
- The computer vision tool also uses machine learning to continuously learn new document types and templates, ensuring continuous service improvement and customer satisfaction.

A New Zealand Insurer is using imagination and combinatory innovation to address persistent processing delays that were creating potential fraud, compliance and brand exposure, and impacting customer experience.

- Massive volumes of returned customer mail, marked 'Gone No Address' (GNA), were presenting the firm with a huge challenge, averaging 150,000 items annually. Compliance rules require those customer accounts to be frozen to prevent fraud. Manual processing was time-consuming and plagued by delays, resulting in negative customer experiences.

- Connected-RPA digital workers, customer contact tools and computer vision were integrated by the firm to swiftly freeze GNA accounts, reach out to account owners requesting corrections, and update new addresses across all relevant internal systems.
- Customer experience was greatly improved by this process, ensuring accounts were available, avoiding unpleasant surprises and service denials, and directly supporting Financial Crime Compliance requirements.

A US tax-preparer applied imagination and combinatory innovation to ensure preparers and customers are ready for service during the seasonal US tax calendar. Staff workloads and customer stress increase exponentially as tax deadlines loom, so time and responsiveness are key factors in creating a positive customer experience.

- The firm integrates Connected-RPA with Microsoft Computer Vision and Regex tools so digital workers can read, extract, and record credentialing and licensing data for its 80,000 tax advisers. Certificates from multiple continuing education providers arrive in many different formats, but the new intelligent digital workers on the case, ensuring staff are credentialed and available for revenue-producing work when tax season opens.
- On the customer side, integration of Connected RPA and mailer tools prompts customers to re-set their passwords in advance of tax deadlines, eliminating last-minute help desk calls, delays and frustration.

A European bank used imagination to beat the clock when faced with an 'impossible' Anti-Money-Laundering (AML) compliance deadline. The bank integrated Connected-RPA with analytics tools and multi-channel web and customer contact systems to refresh, cleanse and update nearly 5 million customer records.

- In a 'Refresh' workstream, the bank trained digital workers to verify and update proof of identity documentation for 3.5 million long-standing customers across multiple bank systems. In a separate 'Cleanse' workstream, the digital workers identified 1 + million customer records that had data quality issues.
- The digital workers then used multiple contact tools—letters, e-mails, and SMS/texts—to request updated documentation from approximately

500,000 customers; upon receipt, the digital workers matched returned data with outreach correspondence and sorted documents into customer files for human review.

- To expedite completion, the bank also built a custom load balancer to optimise workflows across the digital workforce in real time.
- The digital workforce's average daily workload during the project was equivalent to 600 FTEs, with peaks to 850 + FTE equivalents on some days—an impossible manual task given deadlines.

A multinational, multi-line financial services provider used imagination and combinatory innovation to improve customer service and manage cybersecurity.

- Integrating Connected-RPA digital workers with Digital Virtual Assistants and Natural Language Processing tools, the new solution processes customer requests 100 percent digitally, interpreting and confirming the customer's request, scheduling and executing the required work, and confirming completion. Handling times have shrunk from 7 minutes to 45 seconds, creating a transformational customer experience.
- The company also uses digital workers to enhance employee experience and efficiency by automating Enterprise Identity and Access Management, handling 1.7 million employee requests per year for access provisioning on dozens of internal and external applications, tools and data resources required to execute work.

Conclusion

As these examples demonstrate, the financial services industry is once again at the forefront of combinatorial innovation, applying imagination and new technology to shrink time and create 'triple win' value—for customers, employees and shareholders. Perhaps more important, however, on the supply side application and partner ecosystems have been speeding up the innovation process itself. These networks offer rapid, flexible access to a growing family of intelligent tools, applications and services with proven integration pedigrees, thus accelerating time to value and delighting customers. Time really is money, after all.

5

RPA: Mastering Agility at Scale

Scaling is the most important yet most hidden and rarely discussed attribute—without understanding it one cannot possibly understand the world.
Nassim Nicholas Taleb

Introduction

In the previous chapter, we highlighted how financial services companies are applying imagination and combinatory innovation by shrinking the time lag between request and fulfilment, creating a 'triple win'—for customers, employees, and shareholders. Sometimes, however, the business challenge isn't just about time, but size and scale—operating with agility and resilience. How can RPA with other technologies help here?

Scale: For ... and Against

The advantages of achieving and operating at scale, in both private and public domains, are considerable and well established—greater economies of production and service, preferred access to resources, markets, skills, relationships and infrastructure—benefits unavailable to smaller players. Simply put, size matters, or as one revolutionary famously said, *"quantity has a quality all its own."* Despite these obvious advantages, however, operating at scale can

L. P. Willcocks et al., *Maximizing Value with Automation and Digital Transformation*, https://doi.org/10.1007/978-3-031-46569-7_5

also lead to inefficiencies, poor communications, hidden expenses and costly delays.

Large organisations with multiple constituencies are particularly vulnerable, especially those providing products and services we rely on in our daily lives. Legacy systems, multiple handoffs and broken processes can prove costly. In this chapter we look at leading examples of very large companies, service providers, and government entities that are successfully applying imagination to re-invent their operating processes at scale. What does combinatory innovation look like at scale, and what can we learn from their experience?

Multi-sector Examples

We begin with a 100+ year-old US electric utility, serving over 2 million customers. One of the biggest challenges for electric utilities (aside from downed power lines) is customer identity theft—service that's fraudulently obtained using another person's identity. This not only results in lost revenue but also customer frustration with ensuing billing and credit impacts.

This utility initially applied imagination by using analytics to proactively detect potential identity theft based on key indicators—odd consumption patterns, short duration in premises, and account arrears. Given the scale of its operations, however, fraud investigators just couldn't handle the volume of potential theft cases, so the company built an automated screening process using RPA+ machine learning to identify and target the most likely cases from for human investigation. With near 100 percent accuracy, the combinatory solution saved this utility over $3 million in its first year, a sum that will grow steadily as the RPA/machine learning process continually improves, while minimising inconvenience and frustration for identity theft victims.

A UK online retailer faced a similar situation when goods were ordered fraudulently under existing customer accounts. The company used imagination and combinatory innovation to develop an impressive solution. The challenge was massively complex, given that multiple orders might be placed fraudulently on each account, and each order might contain several items sourced from multiple fulfilment centres and delivered by different courier services. As with the US electric utility, the fraud was directly impacting revenue and creating negative customer experiences.

While the complexity was intimidating, the retailer had extensive prior experience with automation, and applied imagination to integrate connected-RPA with analytics tools and teams to identify fraudulent orders, and intercept goods before delivery. Because each courier service used different communications methods, the company built 5 different mechanisms to interface with 3rd party portals or e-mail messaging services to stop couriers from delivering the goods and return them to the retailer's fulfilment centres. The key to driving value was the speed and accuracy with which the connected-RPA digital workers could contact couriers identified for each item.

Combinatory innovation has also re-defined service delivery for a North American provincial financial services company with 750,000 customers. The bank applied imagination and creativity to streamline its end-to-end customer experience while improving its data accuracy and quality. The bank has been using connected-RPA to integrate a blockchain-based platform with natural language processing tools and artificial intelligence engines to transform the customer experience at scale, completely digitising its online customer interface and enabling customers to select and set up their desired services directly.

With 24/7 availability, the new automated processes guide customers through each request using natural language processing to identify the customer request, capture, structure, and validate the necessary data from the customer, automatically accessing multiple internal and third-party resources. The integrated solution creates dynamically configurable workflows that guide customers across three high-volume processes: Customer Onboarding and Account Origination, Account Servicing, and Shared Services. This innovation has dramatically reduced turnaround time for customers—up to 99 percent improvement—while greatly improving data accuracy and quality. The new solution handled over 90 percent of transactions in the targeted processes in the first 12 months of operation with no manual intervention.

For the global pharmaceutical industry, the challenge is all about ensuring timely approval and availability of vital medicines for billions of global customers. The process of developing and delivering new medicines—from laboratory to patient—is complex and challenging, involving development and distribution on a global scale. A global pharma company applied imagination and combinatory innovation at key early- and late-stage points in the end-to-end process.

On the front end, the critical process for gathering and structuring data from the company's clinical trials for regulatory review involves creation of more than 10,000 highly detailed data visualisations annually for Clinical Study Reports. With each visualisation a painstaking task, requiring expert supervision throughout, the process was vulnerable to corrections and rework. The pharma company used connected-RPA, computer vision and analytics tools to read, verify, and structure the raw clinical data. The digital workers then invoked multiple graphic tools to and render the data visually, with tables, charts, graphics, pictures. The entire automated process is completed in less than 24 hours, greatly reducing cycle times for clinical drug approvals.

The 'last mile' challenge in pharma, of course, is getting medicines to patients quickly and reliably. Here the company again applied imagination to build a solution based on connected-RPA. The digital workers now use computer vision to convert import documentation from email submissions into PDF documents, extract key data fields, apply rulesets to highlight discrepancies in the documents and notify local teams of any mismatches or gaps.

The solution improved accuracy and greatly increased employee productivity, saving some 220,000 hours of manual checking when deployed in all markets. Ultimately, of course, as with faster clinical approvals, this innovation enables the company to get medicines in the hands of doctors and patients faster.

On a lighter note, we all enjoy watching TV programs and going to the movies, but we take for granted the global systems and processes that enable us to watch and listen to entertainment—not just production and broadcasting, but the 'invisible' business infrastructure of rights, royalties and distribution that supports and enables the entertainment industry. Applying imagination and combinatory innovation proved the winning approach for a global media services provider in managing multiple intellectual property revenue streams and large media players.

The diversity of revenue streams (ad sales, movie royalties, etc.), spanning three continents, each with global variations, was beset by high-volume, non-standard, data intensive processes, and by specialised systems supporting each revenue stream. The complexity inherent in processing thousands of document variations was overwhelming, and highly vulnerable to errors and delays.

The company applied imagination and combinatory innovation to re-design and simplify their payment processing, building new end-to-end processes supporting high-volume workstreams. The fundamental solution design integrated connected-RPA with sophisticated document processing tools to extract and validate information from thousands of non-standard sources, screens and data models, and structure it for efficient and accurate processing. The connected-RPA platform was further extended to integrate multiple internal systems and applications, as well as external media-specific and web-based tools involved in the payments process. Taken as a whole, the new solution reduced processing times by 70–90 percent, resulting in faster payments, greater partner satisfaction, and improved regulatory compliance.

Conclusion

As we saw in our earlier analysis of financial services companies applying imagination to compress time, we can see here how very large enterprises and governments are using imagination and creativity to master the challenges of operating at scale—re-designing critical processes, integrating connected-RPA with sophisticated analytics, cognitive applications and cloud-based services to improve their operating performance and customer experience. Their early successes and the rapidly expanding array of new intelligent technologies hold great promise for other industries and customer communities challenged by size and scale.

6

RPA and Managing Complexity

Management always hopes to devise systems that are simple…but often ends up spending vast sums of money to inject requisite variety—which should have been designed into the system in the first place.
Stafford Beer
Make everything as simple as possible, but no simpler.
Albert Einstein

Introduction

In previous chapters we've looked at organisations that are applying imagination and combinatory innovation to gain velocity and to master scale challenges in their operations. In this chapter, we delve into the challenges presented by rising levels of complexity, and explore how leading organisations are using imagination and connected-RPA to align their operations more effectively with their business environments and customer needs.

The Inevitable Growth of Complexity

Organisations typically begin life as simple structures. Complexity arises as growth and environmental factors start to introduce more of everything—more employees, more customers, more offerings, more systems, more interfaces, more data, more competitors, more regulation and oversight.

L. P. Willcocks et al., *Maximizing Value with Automation and Digital Transformation*, https://doi.org/10.1007/978-3-031-46569-7_6

One response to increased complexity is to focus on efficiency—standardising inputs, processes and outputs. And standardisation works well when the variety the organisation's products, services, customers and markets is known and limited. But there's an obvious peril in becoming too simple—think buggy whips and automobiles … Balancing simplicity and variety is a strategic challenge facing every organisation.

A strategic concept known as *Ashby's Law of Requisite Variety* holds that to survive competitively, the degree of environmental complexity an organisation faces must be matched by a corresponding degree of internal complexity or 'variety'—in its ability to model, create options and take actions that can regulate the impacts of environmental factors and events. Achieving the right balance between variety and simplification depends crucially on imagination and innovation. The good news is that our workbench today is rich in new technologies, enabling companies to add digital workers and cognitive automation to their workforce to innovate, navigate complexity and strike the right balance between variety and simplification.

Complexity: Striking the Right Balance

Take for example the highly complicated task of pre-building freight trains. A major continental rail operator applied imagination and creativity to simplify and improve the highly complex and technical process of 'pre-building' trains in railyards—assembling a variety of railcar types (freight cars, flatbeds, shipping containers, wagons, tankers, refrigerated cars, intermodal carriages, etc.) in the proper sequence for multi-stop journeys, with railcars and cargo being dropped and added at each stop along the route.

The challenge was complicated by reliance on proprietary, non-standard applications and systems. The manual process was extremely technical, complex, and time-consuming, and the automated solution needed to be highly scalable and reliable. The company used connected-RPA to integrate multiple custom systems, applications and processes used only in the rail industry. Digital workers intelligently provide the right balance between complexity and simplicity, allowing the company to innovate the way it managed a highly complex process and create a solution that was relatively quick to deploy, highly scalable and, crucially, reliable.

Complexity can develop unexpectedly, as a major telecommunications company discovered when the cost of handling a seemingly simple process grew massively prohibitive. The company sold several million ADSL Livebox wireless routers per year, available to customers of the company's broadband services. Nearly 50 percent of the Liveboxes returned to the stores were not in fact faulty. The cost of servicing these customers with non-faulty Liveboxes represented two percent of total capital expenditure. Returning Liveboxes physically was costly to both customers and the stores. How to enhance the customer experience while rendering processes highly efficient at scale? Enter imagination.

The company wanted to provide fast communication for customers facing trouble, stop non-faulty Liveboxes being returned, while providing proactive issue-and-fix identification for faulty Liveboxes, and accurate implementation of fixes. Complementary technologies were integrated with connected-RPA to meet these complex aims, and the technical design reflected the requisite variety needed. At the front end AwareX provided omni-channel 24 ×7 access for subscribers. Reinfer used natural language programming, and machine learning algorithms to convert all inputs into actionable data, and to establish business rules for the RPA software.

As the execution engine for the process the connected-RPA platform provided 24-hour customer support, feeding actionable data into billing, fulfilment, and automated diagnostics for service and network assurance processes. Meanwhile Proadapt brought the technologies together with its systems integration capabilities. The automated solution provided zero-touch workflows, and enhanced customer experiences in terms of speed, consistent, on-demand service. It also reduced returns by 60 percent and saved the company 65 million euros in the first year of operation.

Automation can also be a service in itself—helping customers ensure continuity of critical business processes. A global service provider applied imagination and combinatory innovation to create a suite of flexible, cost-effective, fully automated run and backup services for its clients. The company's digital workforce solution provides a suite of ISO-certified 24/7 continuously available services globally, hosted in the Azure cloud environment, greatly simplifying the task of staffing and ensuring continuity for its clients.

The company's main Robot-as-a-Service offering is a fully scalable, secure, and duplicable service, ranging from a single robot to hundreds of robots, which can be seamlessly provisioned into customer 'run' environments in 15 minutes. For short-term or time-limited requirements, the company's innovative RPA Pay-per-Minute service charges only for the time the robots actually work in production mode.

An associated Run Management service is like an insurance for customers' business continuity and RPA support requirements, providing automated failure discovery and remediation. Event response time for this service is under 10 minutes, with issue resolution under two hours for critical business processes. Together, these digital workforce innovations are setting a new global standard for the industry.

Finally, imagination and combinatorial innovation can help to manage the difficult point where scale and complexity intersect. A major multi-line US bank suffered from costly and error-prone manual efforts required to process payments, transfers and loans on a massive scale. A tripwire in earlier automation attempts: a lot of the information was not digitised. Leveraging the digital workers, the bank designed out nearly all manual work to create automated end-to-end processes.

The solution involved multiple technologies, with connected-RPA providing the integration glue and transaction engine. The RPA platform seized or created images of incoming instructions from clients and routed them to a computer vision tool, while the further integration of a machine learning capability enabled the bank to interpret the data and generate structured work processing forms based on patterns discovered by machine examination.

The bank was able to shrink the time required to develop such forms from an estimated 400 employee-months to less than three months. And because the resulting data was now digitised, the Blue Prism connected-RPA platform could process the data against business rules and enter it automatically into appropriate systems via APIs or systems user interfaces.

The combinatory solution has saved the bank multiple million dollars annually in staffing expenses while eliminating keystroke errors and improving processing speeds. Equally important, the innovative solution has created an attractive and competitive edge for the bank in winning new customers, resulting in significant business growth—the bank's loan business has doubled since the solution was deployed, and the increased demand has been met by scaling up digital workers rather than adding headcount. Over 90 percent of the transactions now flow through these digital workers without any human entry required.

The common thread, or perhaps we should say the driving force in all these stories of increasing complexity is the exponential growth of data. By some estimates the total volume of data in the world and in the average organisation doubles every two years, creating vast 'data lakes' of unstructured information. Organisations are massively challenged to ingest and structure this data in order to analyse and exploit it for insights.

Conclusion

Many of the combinatory innovations we see at this stage of what has become the intelligent automation journey involve tools designed to help with this challenge—optical character recognition (OCR), natural language processing (NLP), machine learning, analytics, and others. The good news is that cloud-based application ecosystems with drag-and-drop integration, are helping organisations accelerate their digital transformation. Our next chapter will look even further ahead, at how the longer-range intelligent automation journey is evolving.

7

RPA and Combinatorial Innovation

In today's era of volatility, there is no other way but to re-invent. The only sustainable advantage you can have over others is agility, that's it. Because nothing else is sustainable, everything else you create, somebody else will replicate.

Jeff Bezos, Founder, Amazon

Technology is a useful servant, but a dangerous master.
  Christian Lous Lange, Nobel Peace Prize lecture, 1921

Introduction

In Chapter 1 of this book, we highlighted how leaders across regions and industries are using imagination and applying combinatory innovation to address key enterprise challenges. We've seen how they're integrating an ever-growing suite of intelligent automation capabilities via connected-RPA platforms to accelerate value, to master scale, and to manage complexity in their organisations. These achievements are both impressive and diverse. But where is all this innovation heading? Who's leading the way? Let's explore some advanced examples that may offer clues.

© The Author(s), under exclusive license to Springer Nature
Switzerland AG 2024
L. P. Willcocks et al., *Maximizing Value with Automation and Digital Transformation*,
https://doi.org/10.1007/978-3-031-46569-7_7

The Possibilities are Limitless ...

Using imagination and combinatory innovation, a specialist financial services BPO provider developed an inexpensive, fast and attractive 'Rent-a-Robot' service based on a connected-RPA platform. The service enables multi-skilled digital workers to be delivered in the same way staffing agencies deliver temporary human work—with no upfront cost and full maintenance and retraining. The provider built an internal 'robot factory' using highly automated development and maintenance tools and processes to build and deliver digital workers to clients, enabling them to take on additional work and meet demand peaks cost-effectively.

The digital workers arrive on their worksites pre-trained on a wide variety of key financial processes, including fund transfers, cost settlements, insurance handling, collection support, deposits contracts, even complex loan data calculations, application verifications and closings—all on multiple platforms and applications, including SAP, SalesForce.com, Temenos, Uniflow and others. Since the cost of the digital worker is a fraction of the cost of human employee and the Rent-a-Robot pricing model contains no upfront cost, clients capture financial benefits within a month of deployment, in addition to rapid availability of fully trained digital workers.

As the pace of technology-driven change accelerates, a global education services provider is using imagination and combinatory innovation in its quest to become The World's Learning Company, with significant investments in new AI-guided education tooling. Its operational goal is to create an intelligent digital workforce, capable of self-learning and continuous improvement across a range of internal and external services, supporting and working alongside its human professionals. The company has established a cloud-based connected-RPA platform in order to exploit AI and cognitive capabilities. It currently uses the platform to manage rights and royalties' acquisition from a complex network of suppliers for image, audio, text and video content. Integrating connected-RPA and Intelligent Optical Character Recognition is also simplifying and expediting qualification and credential checks, and the company is using API calls with RPA and Python coding for applicant on-boarding validation process. Learning by doing, the company is moving rapidly to its vision of becoming The World's Learning Company.

Combinatory innovation isn't limited to for-profit environments, either, as a major regional public health care provider has shown. The provider is using cloud-based digital workers to automate patient care along assigned clinical treatment pathways, with fewer than 1 percent exceptions requiring manual intervention. Where the manual process involved gathering and printing up to 15 pieces of data for each patient, including scans, clinical tests, visit histories, etc., then scanning them into a single PDF file, digital workers now read the content, extract the reason for the referral, and retrieve, merge and upload all the required data using secure, smartcard technology, and alert consultants the file is ready for review. The digital workers actively monitor a caseload averaging of 2000 referrals per week on a 24/7 basis, reducing the time required for processing from 25 to 5 minutes.

Based on this success, the provider has also automated maternity patient self-referrals using connected-RPA and cognitive e-form technology to receive and verify data, register the patient, assign appointments, and confirm details to the patient. Letters from clinics to patients, moreover, will be automatically translated into their primary language and published on the provider's patient portal.

Finally, the provider is building a new combinatory automation for clinical coding for ophthalmology and endoscopy patients that uses connected-RPA and cognitive tools to analyse unstructured data, identify common themes, categorise the data and recommend the likely coding output for treatment records.

These three cases are just illustrative, as the possibilities are in fact limitless. We have seen a major insurance company use machine learning and visual processing with connected-RPA to reduce the time required to assess an accident claim from 56 minutes to five seconds. The early pilot alone was saving five million dollars a year and freeing up 39,000 hours of work time. Another insurance company combined Reinfer cognitive automation with connected-RPA to automate product and sales intelligence and optimisation, management information, and to carry out rapid claims' analytics, while transforming the customer experience. A major US city government is experimenting with using licence number plates as the entry point for analysing traffic delays and flows. Connected-RPA here is integrating Google vision and Microsoft cognitive tools to arrive at dynamic pattern recognition and traffic solutions. One Centre of Excellence executive summarised it for us this way: *"The scope is becoming almost infinite. I keep thinking there are things we can't do then realise a bit later that we can."*

Heading for the Future

Where are all these innovations taking us? We talked to Chieng Moua, resident futurist at a major technology supplier. He sees a revolutionary future, one in which we'll enjoy the capability to download **knowledge as a service** to any digital worker for any requirement. In Chieng's vision, Blue Prism's goal is to extend its customers' ability to apply imagination and combinatorial innovation, built around connected-RPA.

"Maybe customers don't know what these new enhancements should be," he says, *"because they've been running a bunch of old processes. But in the future, as companies have gone out there and scoured the planet and identified the best business practices in retail, for example, customers who want to be a part of that can just pay an incremental fee to download that knowledge as a service. And as the digital workers continuously learn, they'll actually enhance it."*

"What I'm quickly discovering," Chieng says, *"is that companies are still applying mundane work processes to support highly intensive, highly focused critical decision-making events, because they don't want make a bad decision. You can give a CEO the best PowerPoint dashboard with all the data correlated with it and it tells a great story, but this CEO still makes a gut decision. Why is that? Why can't we say, 'Mr. CEO, you may not be making the best decision. Here's three or four or five scenarios. We can run it through our predictive analysis tools. We can actually use our digital talents to do a 'what if' scenario to see if it's actually applicable to reaching our goals—financial goals, quality of service goals—whatever the relevant metrics are.'"*

"That analysis capability in itself will be the new PowerPoint," says Chieng. *"I call it 'creative intelligence'—we can pilot and test any thoughts, information, experiences creatively, taking all these different pieces and moulding them into something new. I call it creative intelligence because it does take a bit of a creative mindset, a natural curiosity, to say 'what happens if I mix this with that?'"*

Chieng's 'knowledge as a service' vision is a practical solution to the formidable challenges presented by our exponentially-increasing store of knowledge—continually rendering much of it outdated. As Bell Labs guru Bob Lucky observes, *"purging obsolete knowledge is probably insufficient in itself to make room for the new stuff ... The complexity of our work is always increasing, similar to the increase in entropy decreed by the second law of thermodynamics ...You don't have to know everything yourself, only how to ask the questions."*

Finally, award-winning economist Brian Arthur argues that technology undergoes its own evolution, similar to biological life forms. His insight anticipates the connected-RPA vision—a future in which entrepreneurs will, in his words, *"combine technology built for efficiency, speed and cost reduction with emerging technologies like Artificial Intelligence and blockchain to enhance empathy and connection."* Increasingly, Arthur believes, these technologies will become more context-aware, more conscious and more human-centric, ultimately giving rise to what he calls 'Conscious Combinatorial Technology.'

Conclusion

These future visions are, of course, only acts of imagination, and there are many other possibilities. It is said that what gets measured gets managed. As we've seen in these chapters, with connected-RPA, what can be imagined gets enabled. The examples we've featured suggest we're in the early stages of what will be an increasingly dynamic and powerful process, and we can expect to see an abundance of new technology 'life forms' in the coming years. At the same time, we do have to proceed much more cautiously, or perhaps a better word is: responsibly. Imagination must be applied not just to what technology can do, but also to what its consequences might be if designed and utilised for unethical and socially irresponsible purposes. The greater the potential impact of the technology, the more this becomes a major challenge.

Part II

Intelligent Automation and AI

8

Intelligent Automation/AI: The Value Potential

An enormous amount of business value is being left on the table…. industrial age metrics may persist as a way of thinking about costs and benefits, but the effects of automation and digital technologies in combination and as platforms are systemic and exponential.

The Authors

Introduction

Research at Knowledge Capital Partners has looked at the strategic use of intelligent automation/AI. We studied in particular automation leaders in five major sectors—banking, insurance, telecommunications, healthcare and utilities, and these sectors are covered in later chapters. However, the research, taken overall, has important implications for senior executives in any sector, if they are to secure the immense business value that can be released through the effective use of intelligent automation and AI. Here we report on those findings.

L. P. Willcocks et al., *Maximizing Value with Automation and Digital Transformation*, https://doi.org/10.1007/978-3-031-46569-7_8

Intelligent Automation and AI: Are You Missing Value?

A top-level discovery is that there is indeed an immense amount of business value available. The surprising corollary is that a vast amount of it is being left on the table. What seems to be happening? Practitioners will be all too aware of what we call the *technology-exploitation gap*—technology advances faster than our capacity to use it. This is a perennial, fundamental phenomenon. But for the present it is enough to note that RPA, intelligent automation—combining RPA with machine learning, NLP, image processing, algorithmic usage, and AI (getting computers to do what human minds can do) are, in terms of general business use, astonishingly way behind where the media and hype like to suggest. All this is understandable—it takes time to develop, implement and institutionalise then exploit these technologies. But what can accelerate this process?

Part of the challenge is trying to understand the nature of business value that can be generated from applying these technologies. Executives are familiar with efficiency gains from technology, and using return on investment (ROI) and total cost of ownership (TCO) criteria to direct their technology investment. You need a different kind of thinking, however, to get to the exponential business value that can be unlocked. Because, as Fig. 8.1 shows, efficiency is not the end of the game at all, but only the beginning.

Figure 8.1 is repeated for the reader's convenience from Chapter 2. In particular we want to focus on the client value orientation path. Many clients seem to evolve from lower to higher value orientations, partly in tandem with their growing experience and understanding of what the technologies can

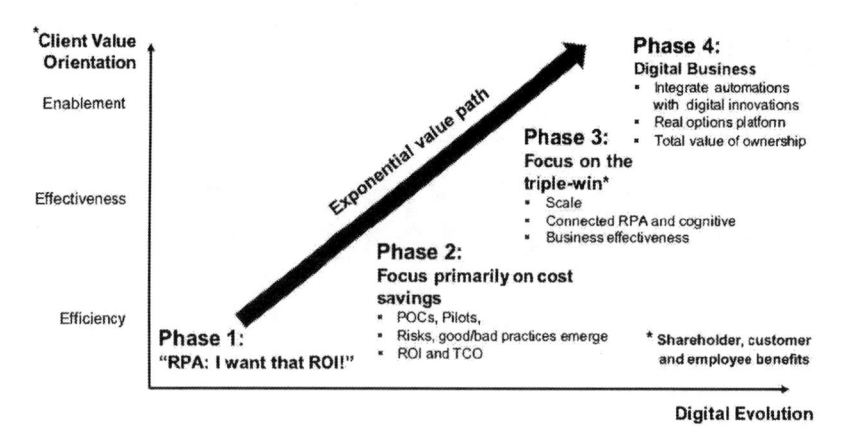

Fig. 8.1 The automation path to value

do. With these technologies business value is coming in three buckets. With each, the kind of value is different, and the business value increases exponentially. Figure 8.1 maps the evolution path some organisations seem to be taking with intelligent automation and AI. Starting with **Efficiency** objectives, they become more convinced by the technology, invest more, then aim for **Effectiveness**, then **Enablement** gains. Other organisations start with a much more strategic view looking for the end–point of creating an automation, then a digital exchange platform that provides **Enablement** gains. Other organisations, so far, have not got beyond the focus on **Efficiency**, and/or do not see automation as a strategic investment. Let's look in more detail at ten major findings to date.

A Progress Report—Major Findings

1. Our first finding is that the potential of robotic process automation, even as a stand-alone technology, is vast and largely unexploited. Our studies of early adopters found examples of ROI between 30–200 percent in the first 18 months, and of 'triple win' shareholder, customer and employee benefits, many unanticipated. Yet, as we saw in Chapter 1, by 2022 the global combined market size was US$136.55 billion in 2022, though expected to experience a compound annual growth rate (CAGR) of 37.3 percent from 2023 to 2030. Compared to the overall global IT market of US$8,852.41 billion in 2023 with a CAGR of 8.2 percent from 2023–2030, and contrasted with the high AI rhetoric, this is a surprisingly low level of overall expenditure.

2. Looking just at the 80 plus major 'connected-RPA' (RPA plus intelligent automation) suppliers, most clients had, by late 2022, between 1 and 80 RPA 'robots' (licences). Few (25 percent) had scaled to 81–100, let alone a higher number. This has been changing in the last year, but reflects scaling, strategic investment and benefits aspiration challenges.

3. Shifting into 2023, adoption was definitely accelerating, but moves into intelligent automation/AI were mainly piecemeal. The main foci were unstructured data capture, business analytics, and improving customer experiences. The economic downturns and slow markets tended to push technology investments into supporting today's business, and dealing with more short-term pain points.

4. The market was still seeing much 'RPA washing': overstating capabilities; overselling RPA and other automation technologies as 'AI', causing confusion.

5. Please refer to Fig. 8.1. The typical organisation gets caught in Phase 1. Initial outlays are small. Frequently the **Efficiency** returns are good, but further investment looks expensive, and the benefits look less clear. With some vendor products enterprise RPA is harder to achieve technologically, and maintenance and support is costly. Looking at traditional ROI-based business cases senior management under-invest, still seeing RPA, and indeed intelligent automation, as tactical tools. Digital transformation efforts may be ongoing but do not connect up, being driven from a different place, with different budgets, and usually under-written from higher up in the organisation. More on this 'blind spot' in a later chapter.

6. A minority of organisations manage RPA and intelligent automation more imaginatively. Experiencing triple win benefits, they extend RPA use to enterprise level, and for back-office, mid-office and customer facing activities. They grasp the potential of 'connected RPA' and 'intelligent automation' and integrate RPA with more advanced cognitive technologies that can manage, for example, unstructured data, analytics, and probabilistic decision-making. RPA becomes the essential execution engine. In earlier chapters we reported many examples of innovative, business changing uses of RPA. Usage included **Efficiency** ('doing things better') but the focus moved to **Effectiveness** ('doing the right things') where much greater value was found.

7. Our research re-discovered the proposition mentioned in Chapter 1—that for RPA, intelligent automation and AI, as for previous technologies, only 25 percent of the challenges are technological and 75 percent are managerial and organisational. This helps to explain the slow progress across Phases 1, 2 and 3. As discussed in Chapter 1, our four previous studies showed 41 material risks arising when trying to introduce automation, as well as established 39 management actions. These actions not only mitigate those risks, but also lead on to effective business deployment. To put it bluntly, organisations that operate in late Phase 2 and Phase 3 became good at managing risks and operating effective organisation practices.

8. The most recent research established also why Phases 2 and 3 are so difficult and why only 15–20 percent of organisations are doing well with digital transformation; and why, depending on sector and definition, 65–75 percent of digital transformation projects fail. Once again, the challenges are mainly managerial/organisational, though getting a variety of emerging digital technologies integrated, deployed and institutionalised is a long haul. It could take most organisations today more

than five years to become digital businesses. We will look at ways forward in Sect. 3 of this book.

9. Looking across our case studies, surveys, interviews and advisory work, we came to a stark conclusion: with RPA, cognitive and AI technologies, an enormous amount of business value is being left on the table. A great deal more value (at least 200 percent) could be extracted by hunkering down and applying RPA etc.much more widely for efficiency purposes. Still more value (the initial indications are 500 percent or more) could be gained by 'hunkering down' and looking for applications that gave business **Effectiveness**. But our latest research suggests that the real value bonanza is in building a digital platform with automation and other technologies, that gives to the business flexibility, adaptability, strategic options and resilience at low cost. Initial indications here are that the value gained is exponential, exceeding **Efficiency** gains alone by ten times (1000 percent) or more. 'Hunkering down' is a profitable fork to take; going big gets you much further, more quickly.

10. This made us ask: Why is so much value being left on the table? What is noticeable and distinctive about those who 'go big' is that they have senior executives who see digital technologies as strategic and transformative; they give sustained support and resources for long-term organisational change using disruptive technologies; they appoint credible influential champions to make this happen; technologies, including automation are seen not as tools but forming a digital platform for new business models, innovation, lines of business and relationships with customers. Interestingly, they rely on 'big bets' thinking fuelled by a big picture view as to what the business needs and the technology can enable, and have less time for cost–benefit and total cost of ownership analyses. By contrast those 'hunkering down' tend to be much more driven by such carefully calculated business cases, and have a much more bottom-up approach to utilising automation technologies for business value. There needed to be a much more convincing way to assess the value available for all parties.

Navigation: The Total Value of Ownership Framework.

This led us to ask the further question: How can we get senior executives to understand the massive potential value that is there, and how to grasp it? In response we developed a Total Value of Ownership framework, shown in Fig. 8.2.

The reason for using TVO is to ensure that business cases for automation and digital technologies are driven by total costs, multiple expected business benefits and by the strategic returns from future business and technical options made possible by these technologies. TCO is arrived at by summing all resource costs encountered across the activities comprising the technology life-cycle. This flushes out hidden costs so often missed when using more traditional metrics. On the value side, we have already found strong empirical evidence amongst digital leaders for a triple win for shareholders, customers and employees. Our three E's framework is designed to capture all these, but also locates further hidden value frequently omitted from client business cases.

For example, much hidden value resides in the potential from business analytics for **Efficiency**, **Effectiveness** and **Enablement**. Additional hidden value is located in the '**Effectiveness**' area (doing things right/differently),

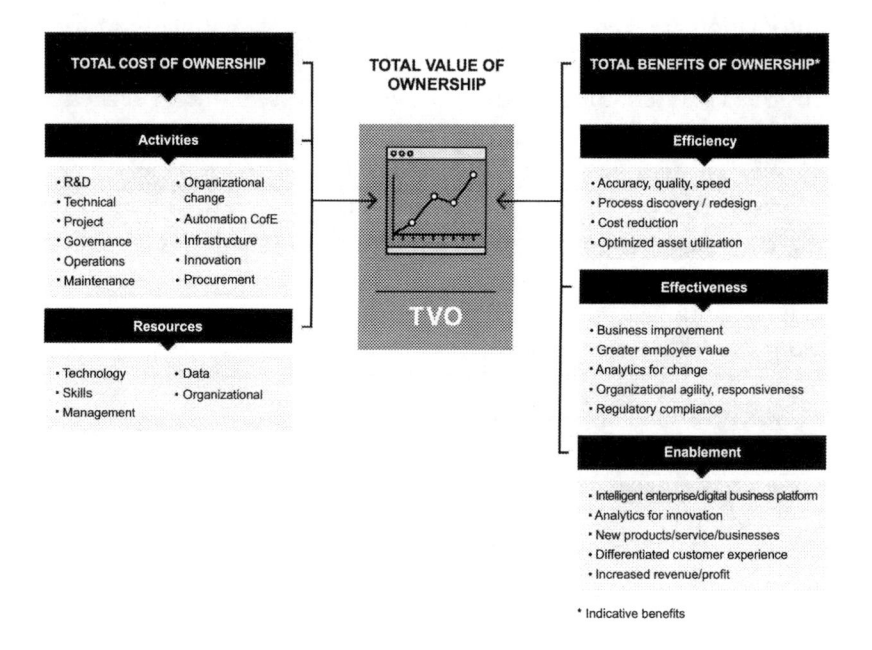

Fig. 8.2 A total value of ownership framework

by using automation to change how business is done or extending business capabilities. Meanwhile when we come to '**Enablement**' we have already found multiple examples of enhanced customer journeys, new services and increased profit/revenue. However, this is just scratching the potential in these directions, given how the technology is developing. Furthermore, we need to capture the hidden strategic value of the future options created where these technologies contribute to building a digital business platform, multi-faceted in its internal and external usage (see Sect. 3). Discounting such major hidden future value is a serious mistake.

You do not have to go far for exemplars that support strategic TVO thinking as a compelling proposition. We give a detailed case of DBS bank in Chapter 24. Amazon is another high-profile example. Starting on-line in the 1990s with books, it re-imagined the customer experience then evolved the platform they built to sell a vast range of goods. Its automation of business processes now ranges from one-click ordering, which connects up the vendors who market their wares on the Amazon platform, the payment systems, order tracking, and optimising the choices for fulfilment. The platform has led to further developments. For example, acquiring Whole Foods, Amazon now operates in the grocery market. A customer can have a Whole Foods QR code on an iPhone to flash at checkout that adds a further discount based on what has been purchased. Amazon has also played with eliminating checkout altogether, using a smart device that automatically tallies purchases when placed in the shopping basket. When finished the customer simply walks out of the store. These are the kinds of '**Enablement**' benefits and opportunities that come from having a robust digital automated operations platform, and which the TVO framework is designed to capture.

Conclusion

Clearly if there is a lot missed business value, there are also multiple ways that a lot more value can be captured, as demonstrated by an admittedly small percentage of organisations in each business sector. Interestingly amongst our cases, even those using combinatory, innovative and impactful use of these technologies did not score themselves highly on measurement and steering techniques. Managerial judgement was outclassing ROI and TCO metrics in these cases, but that cannot be the recipe going forward for the majority of organisations facing rapidly emerging, serious investment questions on RPA, intelligent automation, AI and digital technologies. Industrial age metrics may persist as a way of thinking about costs and benefits, but the effects of automation and digital technologies in combination and as platforms are systemic and exponential.

9

Intelligent Automation in Banking

"Possibilities do not merely add up; they multiply."
Paul Romer, co-recipient, 2018.
Nobel Memorial Prize in Economic Sciences.

Introduction

During 2023, financial institutions were looking to boost by at least ten percent their investment in a variety of digital services, including mobile banking and asset management applications and online trading. Legacy banks were seeking to become more competitive with newer rivals and fintech banks through addressing the online customer experience. Intelligent automation/AI has impacted financial services in ways that include powering recommendation engines and chatbots that aid customer service. AI also helps banks with fraud analysis and prevention. In 2023 some 28 percent were investing in conversational AI, and 31 percent in fraud detection. At the other end of automation, financial services were increasing investments in low-code, no-code solutions to reduce the in-house software development workload, and allow reuse of previous work. Meanwhile more banks were becoming cloud-first, with some 58 percent citing migrating workloads to the cloud as a challenge, not least navigating privacy regulations across different global regions.

Furthering these developments, automation technologies could contribute an additional $US1 trillion annually in value across the global banking

L. P. Willcocks et al., *Maximizing Value with Automation and Digital Transformation*, https://doi.org/10.1007/978-3-031-46569-7_9

sector—through increased sales, cost reduction and new or unrealised opportunities. But this value is still largely being left on the table. Why? As we have suggested throughout, there are well documented challenges with automation, including lack of clear and strategic intent and senior executive support for automation, plus heavily siloed deployment within organisations, resulting in disconnects within and across digital transformation efforts. To be frank, existing operating models per se, mostly neither enable nor ask for strategic use of automation technologies. But a hidden key reason has become increasingly obvious—the failure to grasp the nature and size of the opportunity.

This is understandable. As a species we are not good at understanding the major engines of the opportunity for value creation—compound growth and combinatorial innovation. As our earlier chapters indicated, if automation technology deployment produces small improvements each year, the compound results will be massive. If automation technologies can be recombined in new ways, not only can existing opportunities be seized, but new ones can be created, ad infinitum. As Paul Romer said in 2016:

"Every generation has underestimated the potential for finding new recipes and ideas. We consistently fail to grasp how many ideas remain to be discovered. The difficulty is the same one we have with compounding: possibilities do not merely add up; they multiply."

Prescient banking executives we have been researching understand two things: the strategic opportunities offered by intelligent automation; and how automation can drive the twin engines of compound growth and combinatorial innovation.

They have also, for some time, been anticipating how automation can be deployed to address inescapable competitive pressures driven by rising customer expectations on digital banking. Financial institutions are aggressively deploying automation technologies. This has accelerated during the COVID-19 crisis. Moreover, digital ecosystems are disintermediating and re-shaping how financial services are discovered, assessed, purchased and delivered. Think, for example, of mobile devices, and multi-channel apps made available by fin-tech businesses such as Wise (formerly Transferwise) that have transformed the customer experience, beginning with foreign exchange and expanding into international banking. High-tech multinationals are also entering financial services, leveraging seamless multi-channel customer relationships, advanced scaled technology infrastructures, and immense real-time data lakes.

In all this, intelligent automation has become vital for future competitiveness and differentiation in financial services. Let's look at three banks that have grasped both the problem and the opportunity.

Banking—Grasping the Problem and the Opportunity

Case 1: North America—Human and Digital Workers Blend and Multiply Outcomes

In 2015, a major Canadian bank adopted a new value-oriented, purpose-driven management philosophy of increasing organisational agility and improving customer experiences. A key focus involved transforming disjointed operating processes on an end-to-end basis but from the customer's perspective. This went far beyond simply tweaking existing systems and processes for incremental improvement and cost reduction.

Accordingly, the automation business case was based on increasing the value of the bank's services as measured by customer metrics—retention rates, service expansion, and improved net promoter scores—rather than simply 'doing (bad) things faster'. Taking an agile approach, aided by design thinking, the bank realised that a unified customer data structure was a critical requirement for improving service experience. They integrated front-end artificial intelligence and machine learning tools with their Blue Prism platform to capture, structure, and curate existing customer data in a shared repository supporting multiple service lines.

In addition to **Efficiency** savings estimated at more than 200 percent from the ability to access and use previously trapped data, the bank also estimated a 400 percent gain in enterprise **Effectiveness**—measured by increased customer retention and revenues from broader services integration. The technology platform, moreover, enabled a new organisational structure built on a blended human and digital workforce that could better match task times and volumes to appropriate resources. As the bank's automation lead notes, *"it changes how you think about 'work'."* Taken in aggregate, the bank's gains in **Efficiency** and **Effectiveness** feed and reinforce each other, exemplifying Romer's maxim that innovations *"do not merely add up, they multiply."*

The bank's intelligent automation platform has also supported greater **Enablement** gains in terms of new products and services, enterprise resilience, and first-mover advantage. When the COVID-19 pandemic required major government response, for example, the bank was able to develop custom automations in just a few days to support massive government referral and aid programs. Without adding headcount, the bank was able to complete thousands of aid applications, attracting new customers and generating widespread public goodwill and reputational equity. The bank estimates the resulting gains in enterprise **Enablement** to be greater even than the combined **Efficiency** and **Effectiveness** gains.

Case 2—The Middle East: A Bank Transforms Customer Experience

A major Middle Eastern bank similarly undertook an enterprise-wide transformation to seize a leadership position in its key markets, using Blue Prism technology as a strategic platform, beginning with greater **Efficiency** and **Effectiveness** gains. When customers requested payment investigations, employees had to manually check payment status and customer details to respond to any query. The process involved accessing multiple systems and e-mailing multiple departments—a complicated, time-consuming, and error-prone exercise resulting in multiple e-mail threads that created delays and errors.

By combining intelligent automation with Natural Language Processing (NLP), Machine Learning (ML), and data mining tools, the bank developed a totally automated end-to-end solution to track payment status, pull relevant payment and customer details, and apply rule-based validations, referrals and query responses, with no manual intervention. The solution delivered 100 percent improvement in quality, response times, and customer experience. Employees also gained valuable new skills and experience implementing the automation program. The new process also generated **Enablement** gains from the resulting wealth of data and management information—raw material for applying data science using Hadoop to improve management decision-making.

The bank also gained significant **Enablement** value from automation in responding to the COVID-19 pandemic. At the onset, the bank had to reorganise the working model and infrastructure immediately to accommodate remote working, and simultaneously deal with an unexpected spike in service volumes requiring additional staff. By exploiting the flexibility of its intelligent automation platform, augmented with NLP, ML, and Big Data capabilities, the bank was able to shift to a work-from-home model and support the expansion of demand smoothly, with no business impact.

Finally, it's worthwhile to note that bank's automations underpinned **Enablement** gains through drastic improvement in transaction enquiry handling turnaround time. Given different countries and time zones the bank deals with, often a SWIFT enquiry response that had originally taken 8–12 hours to close, could now be accomplished in less than 2 minutes, resulting in a transformed customer experience.

Case 3: Europe—Reduced Customer Wait Time from 12 Days to 4 Hours

One of Europe's oldest and largest banks, serving over 10 million customers in multiple countries, realised major gains in service quality, speed to market, and customer experience from its intelligent automation deployments. Over 300 acquisitions led to a complicated operating environment with no core banking system. Intelligent automation, however, enabled the bank to manage operations across legacy estates, using APIs to bridge systems and alleviate problems. The senior automation lead describes its RPA platform as the 'arms and legs' that pull data from systems, and cognitive tools such as ML (machine learning) and OCR (Optical Character Recognition) as the 'brains' that analyse and interpret it. The bank estimates it has achieved a significant 150 percent improvement in overall **Efficiency** from its automations and expects additional gains from process improvements in 2021.

The bank also estimates it has captured an additional 30–50 percent value to date in overall enterprise **Effectiveness**—resulting in higher transaction volumes, better regulatory compliance and improved service quality, availability and timeliness. The automation platform has increased enterprise productivity and brought significant growth in both customer and employee satisfaction. On regulatory compliance, the complicated 'Know Your Customer' (KYC) remediation process is now supported by digital workers and presented in 'dashboard' formats for management decision-making. The result: on time with 100 percent quality. And by integrating chatbots with its Blue Prism automation, the bank's customers can request credit and debit card cancellation and replacement in a single fully-automated transaction.

While the bank had not set out to achieve transformational gains, it is doing just that by progressing an infrastructure platform for innovation. Digital workers take on many roles, for example: chat bots that automate customers' bank statement requests; accountants that read income statements from customers, saving time for their colleagues on the front line; work schedulers that park payments during peak volume time to make maintenance cheaper. The bank has already realised an estimated 30 percent additional **Enablement** value to date from its more than 500 digital workers. They enabled the bank to rapidly develop and deploy processes giving customers access to government pandemic aid and relief funds. Core banking services such as loan commitments—previously taking 12 days—are now provided to customers within four hours—a huge expansion in customer added value. Service is now available at weekends, increasing volumes by five percent. And detailed compliance reports for multiple national and European authorities and jurisdictions are now compiled and formatted by digital workers for human review and approval.

Conclusion

What are we learning from these leaders? Firstly, adopting a strategic executive mind-set in deploying intelligent automation is critical in capturing maximum value. Without a transformative view and an enterprise vision suffusing from the top, the strategic uses of automation for greater **Effectiveness** and **Enablement** are foregone by tactical local initiatives, focused narrowly on what can easily be measured: cost savings and cost avoidance.

Next, leaders in automation deployment start with an external focus on customers and competition, using that perspective to design an end-to-end business process architecture that accelerates digital innovation. By 'seeing the business through the customer's eyes', they use automation to improve every aspect of the customer experience rather than 'doing bad things faster'. Creating value is the primary objective; cost is important, but secondary.

Thirdly, building a robust in-house automation capability creates flexibility and a knowledge base which, with strong governance and disciplined behaviours, forms part of the **Enablement** platform and accelerates strategic uses of automation technologies.

Fourthly, longer term strategic **Effectiveness** and **Enablement** value from intelligent automation far outstrips near-term **Efficiency** gains in the leading deployments we have studied—by multiples ranging from 3× to as much as 7x—demonstrating the value of compound thinking.

Finally, in a rapidly evolving multi-vendor technology environment, choosing an open automation architecture is a critical decision factor.

10

Challenges and Customer Experience in Utilities

"Three management practices are central to automation in utilities: mass customisation, service quality, and customer life-cycle management."
The Authors

Introduction

Few industries have been as buffeted by change over the past 25 years as utilities. Industry restructuring, globalisation, deregulation, and digital technologies have up-ended a traditionally comfortable, almost leisurely operating model: regulated monopoly, vertically integrated supply, predictable demand, long service lifecycles, and highly commoditised products. The scale of public utility operations, however—once a formidable asset and barrier to entry— has become a potentially massive liability in a disaggregated value chain as more agile intermediaries emerge—resellers, digital-first utilities, and alternative sources of supply. During 2022–2023 the ongoing Russia-Ukraine war provided a further large-scale disruption to energy suppliers globally.

Our research shows that utilities have already widely adopted RPA for a range of tasks. These include: verifying meter readings. facilitating customer billing, doing pricing calculations and order entry, updating inventory records, sending notifications related to time-sensitive contract renewals, and offering self-service options for asking common questions, resetting passwords, and setting up accounts. Meanwhile on some 2023 estimates intelligent automation could save the industry US$237–US$813 billion when

L. P. Willcocks et al., *Maximizing Value with Automation and Digital Transformation*, https://doi.org/10.1007/978-3-031-46569-7_10

scaled. How? Research in 2023 suggests that this would not be just from cost savings. Already we are seeing in some organisations quicker access to customer insights and data, faster time to market in launching new products, positive impacts on customer satisfaction and increased customer retention.

When customers in any sector are offered choice, research shows that *service experience* across the customer lifecycle—pricing, speed, responsiveness and reliability—becomes the major determinant in customer acquisition and retention. And the ability of agile challengers to respond to these drivers becomes a critical competitive advantage. At the same time, incumbent providers' massive operating scale—a seeming impediment to agility—also creates a perfect laboratory for innovation.

The business challenge is the same for incumbents and challengers alike: how to offer reliable, personalised service at scale with constant customer-perceived improvements and keen pricing. The strategy adopted by incumbents we've researched is twofold: mass customisation—applying digital technologies like RPA and cognitive tools to match their challengers' service innovation strategies—while continuously controlling the cost base and meeting regulatory compliance. For both incumbents and challengers, however, the solution begins first and foremost with the customer experience, and intelligent automation is a powerful competitive weapon.

Better Service Through Automation

Three management practices are central to automation in utilities: mass customisation, service quality (SERVQUAL), and customer life-cycle management. Mass customisation means applying mass production techniques to what is invisible to the customer—e.g., a car engine—and customisation to what is visible—e.g., seat upholstery. With highly commoditised products, it is in the customer-facing service dimension—where incumbents are most exposed and competition is most intense—that utilities can apply mass customisation most productively.

It is an oft-quoted truism that the customer's perception is the provider's reality. Nowhere is this phenomenon more acute than the potential gap between the customer's service expectations and the actual experience. SERVQUAL is a multidimensional research instrument and gap model across five dimensions of service quality:

1. **Reliability**—the ability to perform the service dependably and accurately—is the key service dimension.

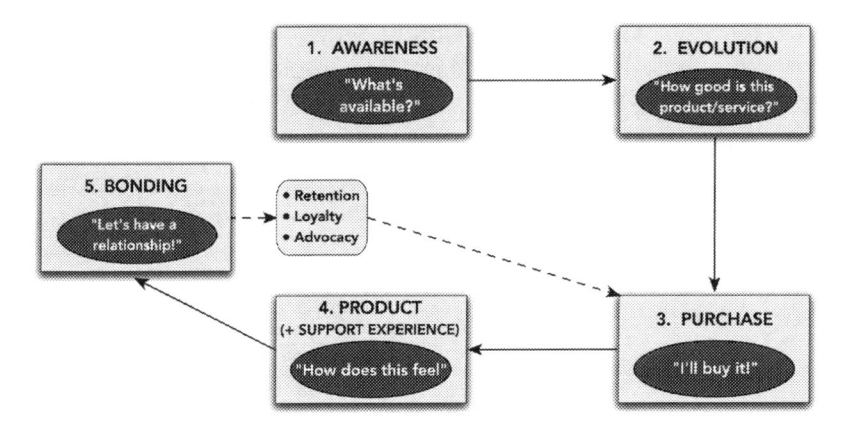

Fig. 10.1 The customer lifecycle

2. **Assurance**—the knowledge and courtesy of staff, and the confidence and trust they generate.
3. **Empathy**—the provision of caring, individualised attention.
4. **Responsiveness**—the willingness to help the customer and provide prompt service.
5. **Tangibles**—location, appearance of staff, call centres, equipment, etc.

Today utilities have to focus on all these dimensions—at every touch point of a customer's experience—to compete effectively. As the Fig. 10.1 shows, these touchpoints are surprisingly many. What is impressive is how intelligent automation can leverage each of these touch points through high service, while generating analytics to improve and anticipate future service. This approach enables granular personalisation for mass customer bases. What is surprising is how few utilities have yet grasped the full potential of these opportunities with automation. Let's look at three utilities that have.

Applications in Utilities

Personalisation at Scale

In 2008, a national provider of electric power to residential and business customers became one of the earliest adopters of process automation, conducting several proof-of-concept implementations to explore the capabilities and potential of the technology. Typically, the provider's initial focus was on **Efficiency**—reducing operating costs by applying technology to handle rising transaction loads rather than hiring additional staff. At its peak, the utility's digital workers were handling 500 separate processes involving 18 million work items, doubling the company's initial forecasted productivity gains as automations grew.

Mirroring our research in other sectors, this initial **Efficiency** focus led on to more significant gains in **Effectiveness**. By automating back-office processes and applying human workers to more complex customer-facing tasks, the utility simultaneously improved both regulatory compliance (traceability and auditability) and customer satisfaction (personalised service). Additional **Effectiveness** gains arose from using digital workers to manage operating workflows, automatically assigning workloads to available digital workers and generating completion reports. By analysing the process data captured from its **Efficiency** gains, moreover, the company reconfigured end-to-end cross-functional processes, generating significant gains in employee satisfaction. The company reports that its overall **Effectiveness** gains were *five times* expected results.

The company's automation platform also delivered critical **Enablement** gains when the COVID-19 pandemic emergency required staff to operate remotely from home, using their own personal technology—a potential nightmare for a highly regulated industry dealing with sensitive customer data. With its intelligent automation platform, however, the company rapidly built and deployed an entirely new process in two weeks, enabling staff to access servers from home using RSA security system tokens. The virtual work solution enabled thousands of staff to support customers during multiple extended lockdowns. The intelligent automation platform also enabled the company to extend emergency credits proactively to pre-pay key and card customers who couldn't access brick-and-mortar stores during lockdowns. This **Enablement** innovation created a uniquely powerful customer experience and widespread goodwill.

Industry Restructuring

Global energy industry restructuring—divestment and re-aggregation across the value chain—is creating multiple demands for intelligent automation solutions. When a major energy conglomerate split its power generation and energy trading operations from its retail operations, it established a new listed company in 2016 with 13,000 employees supporting retail operations in over 40 countries.

The company used RPA to optimise processes, increase compliance, and improve reconciliations in large-scale energy trading activities for greater **Efficiency**. Integrating chatbots and artificial Intelligence tools with its RPA platform, the utility automated its end-to-end Procure-to-Pay process. Digital workers now create 120,000 purchase orders per year, generating new supply contracts and updating existing ones directly in the company's SAP system. They confirm valid certifications from 4000+ suppliers and send renewal requests where required, using an AI algorithm to distinguish certificate types. And they prioritise, bundle and dispatch urgent Purchase Requisitions to the responsible business groups, and book 50,000 goods receipts per year—all automatically.

While wholesale transactions are far fewer than retail in numbers, the cost of an imbalanced energy portfolio is hugely consequential for business **Effectiveness**, given the scale of each transaction. Seconds saved in entering data to trading websites represent millions of dollars in value. Digital workers now make trading checks every 5 minutes to ensure the energy portfolio is balanced across supply and demand, a critical metric for customers' success. Customer satisfaction metrics have increased to 95 percent 'fully satisfied,' while supplier relationship management has greatly improved with faster payments.

Transaction speed is critical in a competitive marketplace, but management also realised that intelligent automation was a powerful **Enablement** platform to instil and sustain a strong performance culture for the new company—built on values of agility, responsiveness, and rapid decision-making. Employee satisfaction and employer branding rank among the most valuable benefits of the automation program. In the company's own words, *"It is change management…not a 'technology' initiative…with a key focus on culture change, organisation transformation, and people enablement."* The company estimates **Enablement** gains in building a 'digital first' culture to be more than twice the combined **Efficiency** and **Effectiveness** gains realised to date.

Scaled Retail Distribution

Our third example—a multi-national retail energy provider with local operations in 30 countries serving 33 million customers—established a shared automation resource group to support its national customer-facing businesses. The automation Centre of Excellence provides a trio of transformational consulting resources that ensure consistency, quality, and integrity across the multi-enterprise business family, creating innovations at scale:

- The Business Process Management team works with operating units to identify and assess the best opportunities and use cases for automation.
- The RPA team, builds, tests and deploys automations delivered to business units on standardised platforms.
- The Intelligent Automation team, that identifies, assesses, and works with the RPA team to integrate task-appropriate cognitive and Machine Learning tools to deliver the required automations.

Each of these resources maintains a library of proven objects and automations, along with experienced and knowledgeable specialists, to ensure quality, promote consistency, and exploit innovations across the wider enterprise and its operating units. The management objectives of this model are:

- **Efficiency**—to reduce transactional outsourcing by bringing work back in-house and to reduce demand loads on staff via self-service solutions.
- **Effectiveness**—to improve regulatory compliance, capture end-to-end analytics across the value chain, and achieve higher transaction throughput across operating units.
- **Enablement**—to improve overall customer experience across retail units through efficient onboarding, contact and complaint resolution.

Efficiency gains are multiplying across its highly distributed operating environment, the company has already doubled its expected **Effectiveness** gains, and is on track to achieve significantly greater **Enablement** value as innovations continue to gain scale across multiple business units.

Digital Challengers

On the challenger side of utilities competition, a host of emerging digital natives are exploiting intelligent automation to build streamlined operations platforms and transform utility customers' experience. Without the burden of maintaining and upgrading massive business systems, they are designing service platforms from scratch that ride atop legacy industry systems and processes, taking a demand-side, 'customer-inwards' approach rather than the traditional supply-side, 'product-outwards' model that shaped incumbents.

The benefits of this approach explicitly align with the key service touchpoints in the SERVQUAL model, from Awareness through Purchase to Bonding. As one CEO put it: *"Customer service is not just about answering the telephone, or an email, but it's about taking a proactive approach…and turning what could often be a negative experience—contacting your energy supplier— into a positive one."* By focusing on automating processes and data flows that might be critical but don't require human interaction if done correctly, challengers enable their people to focus proactively on building positive customer relationships, creating richer experiences for employees as well as customers.

Conclusion

Taken together, these examples demonstrate that the 3Es of our Total Value of Ownership model are mutually reinforcing in the Utilities industry, for incumbents and challengers alike. As competition spurs innovation across the industry—enabled by intelligent automation—the journey path to greater value is clear: **Efficiency** gains feed directly into enterprise **Effectiveness** and customer **Enablement,** as value grows exponentially.

But transformational success requires a distinctive kind of executive leadership. Our research reaffirms earlier findings on IT potential by Michael Earl and David Feeny. 'Believer' executives—unlike 'atheists', 'hypocrites', 'waverers', and 'agnostics'—are not just convinced that intelligent automation and 'going digital' will enable strategic advantage. They demonstrate their beliefs in their decision-making, pursuit of opportunities, resource allocation, problem solving, and daily behaviour.

Ask yourself, what is your vision for intelligent automation? If it is not to irreversibly transform key parts of the business, you are on the dangerous, not-so-yellow-brick-road to deteriorating competitiveness and passing up massive value.

11

Risk and Intelligent Automation in Insurance

> The word 'risk' derives from the early Italian word 'risicare', which means 'to dare.' In this sense, risk is a choice rather than a fate.
>
> **Peter Bernstein, *'Against the Gods'***

Introduction

What will the insurance industry look like in ten years' time? What role will intelligent automation play? What are the choices? Are you preparing? These are critical questions for an industry whose essence is calculating risk. As a data-driven business—from risk assessment and underwriting to distribution, and from pricing to claims—the entire industry value chain is ripe for intelligent automation and digital transformation. How so?

Present technological trends are consistent with the following possible 2030 realities. Insurance purchasing is exponentially faster. Risk profiles are automated and updated in real time. A much wider range of customers receive more or less instant quotes. Blockchain applications enable smart contracts and fast payments. Policies provide micro-coverage via multi-party insurance and adapt dynamically to individual behavioural patterns and needs. Fewer agents rely heavily on technology to carry out many more tasks. Underwriting is automated to a few seconds for most customers across life, property and casualty insurance. Predictive analytics allow pro-active, complex policy offers to customers. Through automation, pricing has become massively sensitive

L. P. Willcocks et al., *Maximizing Value with Automation and Digital Transformation*, https://doi.org/10.1007/978-3-031-46569-7_11

and competitive. Differentiated customer experiences provide the key metric and expectation, but profit margins are very thin.

Claims processing, including fraud detection, is 90 percent automated, using the full array of available technologies. Headcount is 80 percent lower than today, with processing times measured in minutes, even seconds. Pre-emptive technologies, for example internet of things in the home or car, are massively focused on reducing claims before they arise. Except for unusual, contested and complex claims, customer service is largely automated and claims settled within minutes. In the face of all this, not surprisingly, regulation has become highly technologised, focused on reviewing and approving machine learning-based models, data usage, and underwriting practices.

Today's Challenges for Insurance

How does this happen? There are some obvious drivers. Customers' digital expectations have become higher. The pandemic crisis has accelerated the adoption and convergence of automation and digital technologies. It works, it's available, it provides resilience and fall-back. Insurers can see new uses. Furthermore, in very competitive markets, 'insurtech' companies already provide stiff competition. They either take a customer-focused approach and target traditional insurers' pain-points and inefficiencies. Alternatively, they pursue a direct-to-consumer strategy, launching new easy-to-use products, and pressuring adoption of automation and digital technologies.

An even greater driver: the historic and unsustainable high operating costs across the industry. Unlike other large-scale industries such as automotive, telecoms and airlines, large global insurance players (with some leading exceptions) have generally not improved overall productivity in the last ten years. While investments in automation have boosted labour productivity, overall cost ratios have not improved. On this issue alone the industry is ripe for structural changes to business and operating models.

Leaders are already taking action: according to a McKinsey study the top 20 percent take nearly all the industry's economic profit, and are notable for their close cost management. Yes, some are very large companies that capture economies of scale. Others benefit from less complex operating models in highly standardised market segments, e.g., bancassurance risk products, but still others have been heavily investing in digitalisation and automation and are starting to see the benefits.

The challenge, therefore, becomes structural change, not least simplifying the end-to-end business model and capitalising on the massive opportunities provided by digital and automation technologies. During 2023, insurance companies expected to increase their automation expenditure in areas such predictive analytics at scale, omnichannel assistance, digital documentation and the Internet of Things (IoT) applications for loss prevention. A lot of digital investments went into efficiency. In our research into leading organisations, however, it has become clear that solving the insurance industry's efficiency problems, by itself, is an insufficient strategy. It can be seen from our 2030 scenario that the industry, aided and abetted by advanced technologies, will move on from greater efficiency to becoming predictive, then future-ready. As of today, then, the three immediate areas for attention are efficiency, yes, but also innovation and customer experience. Let's look at some illustrative cases.

Applications in Insurance

Case 1: Global

In keeping with our '3Es' journey narrative, this global provider began by capturing **Efficiency** gains, thereby releasing funding and resources to support more advanced transformation programs. With a federated operating framework, the company created a global Centre of Excellence, co-located with its global IT, cloud, Infrastructure, and Business Services teams. The CoE established preferred supplier relationships for RPA platforms and cognitive tools and published guidelines at a global level, enabling local business entities to identify where to focus first, depending on their current operating state.

With overall enterprise architecture responsibility, the CoE provides advisory services to operating units, along with end-to-end 'Automation-as-a-Service' and RPA 'Platform-as-a-Service' delivery capabilities. During the first stage of deployment, the focus was on leveraging the legacy enterprise data and application estate for efficiency and cost reduction. Its larger remit is to help business groups scale automation in a cost-effective manner.

The company reports strong **Efficiency** gains to date from 24/7 operations, supporting higher employee productivity with lower recruitment/ training costs.. Process improvements have seen fewer FTEs utilised, and some staff redeployed to higher value tasks. These gains, in turn, have supported wider business **Effectiveness** from higher and faster transaction throughput, increased asset utilisation and ROI, and workload relief for IT applications developers. Automation has also improved regulatory compliance, strengthened security and produced greater employee stability and job satisfaction.

These gains in **Effectiveness** have, in turn, led on to significant gains in enterprise **Enablement**, including better analytics, development of new products, greater enterprise scaling, responsiveness and agility, resilience in the face of event threats (including the COVID-19 pandemic), differentiated customer experiences achieved relatively cheaply through RPA, and more agile and streamlined end-to-end value chains, all resulting in deeper and wider market penetration and share. The top three gains so far have been in workforce optimisation, customer experiences and resilience, but much more is expected, and the company is well along the path to digital transformation.

Case 2: United Kingdom

A UK insurer initially sought to capture value from what it called 'automation arbitrage'—cost **Efficiencies** across multiple business problems resulting from what the leading Automation executive describes as *"growth through evolution rather than design"* that had led to a very complex organisation.

Beginning in 2016, the initial strategy focused on automating stable processes for cost avoidance, concentrating in particular on reducing the transaction volumes it was sending to its BPO provider. The solution involved disaggregating the customer on-boarding process, applying digital workers to handle complex transactional parts of the process fast and accurately, while relying on the BPO provider to handle parts that required human interpretation. As a further development the introduction of more cognitive automation tools will probably see more of the process being brought back in-house.

The primary focus initially was on automating a defined set of stable processes for greater **Efficiency** and reducing costs—there was a lot of 'low-lying fruit'. Fewer FTEs are being used, and staff are redeployed to higher value work. Automation made inroads into cost avoidance and rework reduction. As an indicator, at one stage 21 robots were doing the work of 50 FTEs, more consistently, and at lower cost. A spin-off overall was improved data quality and integrity. The company aimed for and achieved efficiency gains of some 150 percent.

The automation targets have also moved to realising value. The company has gained greater enterprise **Effectiveness** by applying their automation experiences to improve other processes, creating what it calls an 'evolving business case' as automation has grown organically. The Automation lead commented: *"you automate the 50,000 things you do every day; the things that take up your time."* Automation greatly improved regulatory—mandated compliance, for example—transaction accuracy, traceability and auditability. Automation led to more effective use of the outsourcer, giving them the human work in an otherwise automated system. Workforce augmentation through automation means that large numbers of new customers can be enrolled quickly without disruption, when, before, it took weeks and additional staff. Such improvements increase enterprise ROI around the clock.

Further gains in enterprise **Enablement** came from added capability, for example applying what the company calls AFTEs (Automated Full Time Equivalents) to handle new workloads—work that wasn't being done before or short-term demand spikes—without hiring temporary human workers. Massive gains have been made in increased resilience, and strengthened, rapid administration, especially during the 2020–2021 pandemic, not least in the healthcare parts of the business.

Taken together, the company's automation gains have been remarkable. As mentioned, it fully captured its expected **Efficiency** gains of 150 percent compared to its previous 'run' model cost, but it also estimates an additional 150 percent unplanned value gains from increased enterprise **Effectiveness**, and a striking 450 percent gain in superior enterprise **Enablement**. This insurer exemplifies the pattern we have seen in other sectors: intelligent automation value is exponential, not linear.

Case 3: Europe

This major European insurer started with RPA in 2015 amongst the Life, Commercial and Claims parts of the business. By 2017 it had built an automation CofE, then extended to a federated model, with a hub in Claims. The concept was to help other business lines improve processes relevant to them, then scale to an enterprise connected digital workforce. Customer and governance benefits from automation have been more recent targets.

The insurer had planned 100 percent **Efficiency** gains, and so far, has realised 80 percent. These gains come from 24×7 operations, fewer FTEs needed (e.g., in one country at one point 55 robots ran 125 processes), process improvements, and some cost avoidance. Automation greatly improved speed on payments, checks and response to customers (e.g., checks from 24 hours delay to one hour). Security has also improved.

As with our other two companies, a lot of the **Effectiveness** gains—in this case estimated to be actually 100 percent—were not planned. They come from a mix of higher throughput volumes, increased enterprise ROI, avoiding the IT queue, early links with cognitive tools, optimising skills sets (recent), improved regulatory compliance, whole organisation margin improvement, better quality data for analytics, and improved customer and employee engagement/satisfaction. The company has also experienced big gains managing critical business processes (e.g., disaster recovery), dealing with backlogs, handling process peak periods, and being able to switch across processes (e.g., HR, Finance and Claims).

The company is progressing **Enablement** gains, planning 100 percent, and achieving, so far, 45 percent of these. The company now has an **Enablement** platform which is already producing better analytics for decision-making. The platform has increased disruptive potential and first mover advantages in the marketplace (e.g., faster acquisition and integration of brokers). The intelligent automation platform has facilitated great strides in providing a differentiated total customer experience (e.g., faster claims, better up-to-date information, speed to quote) and offered greater resilience all round during the 2020–2021 pandemic crisis. More processes are in the automation pipeline, together with increases in the digital workforce (in one country from 75 to 120 digital workers), and a scaling of the CofE to 75 staff.

Conclusion

We have seen three insurance companies well on the way in their intelligent automation and digital journeys.

All three evolved their automation, gained early **Efficiency** wins, and discovered new value as they applied automation more knowledgeably. Frequently the **Effectiveness** gains were not even guessed at, let alone planned for. Companies that follow can no longer have that excuse.

Our executive respondents all commented how difficult it was to estimate potential and actual **Effectiveness** and **Enablement** gains. However, all could forcibly state where the gains had been achieved, and *where future gains would arise*. The estimation/measurement problem has always haunted investments in information and communications technologies. This is why we placed the vision of likely 2030 realities at the beginning of this chapter. Can anyone honestly refute what the trends are, how they are accelerating, and what the likely outcomes are likely to be for the insurance industry, worldwide? The three companies we have looked at are finding their way forward. Direction, trial and error, patience, and long-term vision and investment have become the keys in the insurance sector.

We began with a quote on risk. Of course, the insurance sector, understandably, is, culturally, very risk averse. This is absolutely right for certain parts of insurance business. But in other parts, just because risk, cost and gain cannot be precisely calculated, does not mean that risk, cost and gain do not exist, and can be safely ignored. Sometimes the risks of taking action are much smaller than not acting. Intelligent automation and digital transformation are already game changers for insurance sectors worldwide. Are you on course for our 2030 vision, or can it be safely discounted?

12

The Knowledge Switch in Telecommunications

In an economy where the only certainty is uncertainty, the one sure source of lasting competitive advantage is knowledge.
  Ikujiro Nonaka, The Knowledge Creating Company

Introduction

Between 2021 and 2023 the general recognition was that telecom operators must marry operational efficiency and reduced capital expenditure with improved customer experiences, higher revenues, and greater business and operational resilience. Into 2023 we saw many operators using IA (Intelligent Automation) technology to create new customer experiences and automate back-office processes. Others were deploying IA to better utilise 4G and 5G networks, or selling IA as a service to their enterprise customers. A recent development was telcos using IA in conjunction with open application programming interfaces (APIs) and an open digital architecture. This allows them to evolve from being telcos to techcos. The evolution to new revenues, including the monetisation of 5G, needs long-term software-defined, virtualised networks, cloud-based IT operations and open digital architectures. Another important part of these long-term moves, also helping to underpin today's business is IA deployment.

While intelligent automation can be operationalised to support multiple objectives, it necessitates **four knowledge-related change imperatives**:

L. P. Willcocks et al., *Maximizing Value with Automation and Digital Transformation*, https://doi.org/10.1007/978-3-031-46569-7_12

1. Improved data control, focusing on identity management and unifying customer profiles across service lines.
2. Connecting customer journeys by identifying gaps and end-states, enabling flexible journey orchestrations in real time.
3. Telcos need new impetus in organising around the customer, with staff engaged with customers' data and empowered to fulfil customers' needs.
4. To build on the integration and personalisation capacities of intelligent automation, as well as the predictive and interactive capabilities, to automate and orchestrate the actions needed.

For telcos, the pandemic and economic crisis of 2020–2022 accelerated and amplified trends that were already redefining the basis for success in the industry. The telco challenge rests, firstly, with the massive amounts of data and information in their businesses, much of which is fragmented, isolated and, at best, semi-structured. Telcos are like utilities in this sense, but with more diverse product portfolios.

Second is the long-term reengineering challenge for telcos to organise around the customer rather than the service. Operations infrastructures and accompanying data repositories have grown up around specific technologies and service lines (e.g., POTS, wireless, internet, cloud, streaming) as they were invented and deployed or acquired, then managed and regulated. Today, they urgently and fundamentally need a knowledge switch, and this poses a challenge.

The historic focus on operational knowledge needs to be augmented with other data, such as what customers are doing, what they want and need to know, and how self-service can be achieved. Connecting data islands, mining for insights, and getting the relevant information structured and accessible to frontline staff and customers is the key to competing, including against digital start-ups. Fortunately, intelligent automation can be a huge enabler if the strategic imperative is recognised, as the three cases below demonstrate.

Gaining Value: Case Studies

Case 1: North America

One North American telco began its intelligent automation journey in 2015 by establishing a Centre of Excellence (COE) to focus on technology selection, opportunity evaluation and prioritisation based on alignment to business strategy, along with building, testing and deploying automations.

The COE leadership wisely adopted a strategy to 'infect' the organisation and create demand for automation by demonstrating and widely promoting results from early deployments.

The success of the initial focus on **Efficiency** with RPA in the back office improved speed, volume and accuracy of order transactions, payment processing and inventory management. This created 'permission' internally to expand automation into front-office deployments to improve enterprise **Effectiveness**.

In contact centres, the integration of RPA with cognitive tools such as Natural Language Processing (NLP), Optical Character Recognition (OCP) and sentiment analysis enabled digital workers to engage directly as chatbots; they supported agents by gathering relevant data in real time from multiple service lines, accelerating case completions and greatly improving both customer and employee experience. Customer sentiment analysis, and the improved customer experience overall, also created opportunities to upsell at scale. Together, the increase in enterprise capacity has translated into significant gains in enterprise **Effectiveness**. The company now has over 240 digital workers in contact centres and is adding 70–80 per year to meet customer demand.

Consistent with results in other sectors and companies, these gains from intelligent automation in overall **Effectiveness** greatly exceed its gains from **Efficiency**. While the company fully met its **Efficiency** targets through automation, it reports achieving *8 times* those gains in greater enterprise **Effectiveness**—double its expectations.

Finally, as seen in other industries, the COVID-19 pandemic created immediate resilience challenges in operational performance and customer experience. Dramatic spikes in service utilisation and traffic patterns as a result of lockdown restrictions and home working—for customers and staff alike—created critical, unanticipated service challenges. By expanding the digital workforce supporting its contact centres and technical support staff, it was able to accommodate these shifts in customer demand and resource availability. This generated additional **Enablement** gains equal to its **Effectiveness** gains, with even greater value expected in the future.

Case 2: Europe

The first European telco interviewee operates on a federated model with four customer-facing lines of business and a fifth focused on IT. While the company predominately uses ROI for evaluating automation business cases, the extensive use of lower-cost offshore service providers often puts stress on building the strongest cases. One solution has been to allocate 'common costs' of automation onto the COE, making automation cases more attractive.

As seen in other cases, the strong focus on **Efficiency** includes over 450 automated processes, which have enabled the company to significantly reduce overhead costs; in its most recent financial year, reduction of future hiring needs plus greater cost avoidance represent 60 percent of the financial benefit from automation. One innovative automation alone generates millions in annual procurement savings by checking excess inventory stocks across all lines of business before confirming new equipment orders. In total, the company has tripled its planned **Efficiency** gains through automation.

Those achievements are multiplied by additional gains in enterprise **Effectiveness** from intelligent automation, with an astonishing *eightfold increase* over original estimates. In one example, with over 200 applications requiring various levels of security access, the company developed a 'validation digital worker' to confirm access for new joiners, terminate leavers and re-validate contractors supporting over 2000 systems.

In its contact centres, the company has realised significant **Effectiveness** gains by creating a digital workforce to communicate with customers via text; the range of requests includes automated fault identification and reporting, setting service appointments and reminders, booking engineers when required, and closing open tickets. Overall, customers get a better experience through streamlined, more focused communications. On path to achieve or exceed its target traffic reduction in contact centres of 80 percent, a more meaningful benefit is the improved customer experience it creates. A 30- to 45-minute call centre interaction now takes less than 10 minutes, and two-thirds of customers rated the service 8 or above on a 10-point scale, with nearly half scoring the service 10.

Finally, the COVID-19 pandemic proved a massive stress-test in **Enablement**. Lockdowns led to a 50 percent drop in on-shore and off-shore labour capacity. Simultaneously, demand for broadband exploded as customers began working from home. Within two weeks, the existing text-based digital workforce was expanded to meet this demand spike, and has now become a sustaining source of long-term value. The company reports that its gains in **Enablement** to date have reached *20 times* its expectations, demonstrating that the full benefits of intelligent automation are both cumulative and exponential.

Case 3: EMEA

The EMEA multinational telecoms provider sought value in the form of cultural change as well as enterprise performance from its automation program. A 'big bang' digital automation program across business and consumer contact centres built on an RPA platform has transformed both customer and employee experience with impressive results.

In its domestic market, the company began by transforming its business-facing contact centres, which handled 160,000 calls and 360,000 emails annually to carry out 500,000 requests across 3 business segments (small, medium and large enterprises). B2B call centre employees had to interact with more than 70 different screens and more than 30 different systems in many requests, due to disconnected service contracts, products and relationships. The company set an **Efficiency** goal to reduce average processing time by 50 percent and the number of screens by 85 percent with a more agile and simple operation. By applying RPA and machine learning, the company has achieved a 35 percent gain in **Efficiency** from its B2B contact centre automations.

While the main strategy goal of the project was to improve **Efficiency**, the deployment also generates **Effectiveness** gains; with more than 50 percent reduction in average handling times, quality and customer satisfaction have greatly improved. Digital workers freed human workers to focus on higher impact activities, so employee satisfaction also rose significantly by eliminating frustration and allowing employees to focus on customer engagement.

The **Efficiency** and **Effectiveness** gains realised in the B2B contact centre automation program are being replicated at scale through the consumer-facing contact centre transformation program. The B2C contact centres get more than two million calls per year, receive 60,000 emails and carry out four million operations, and automations are expected to reduce agent time more than 30 percent, again improving both employee and customer experience.

Combined, the business-to-business (B2B) and business-to-customer (B2C) contact centres were expected to deploy a total of over 1000 digital workers. They will pave the way for even greater **Enablement** gains as the company experiments with more sophisticated automations, such as cognitive-skilled digital workers capable of interacting autonomously with customers to solve support requests.

Conclusion

The three telecom leaders interviewed responded to the telecom imperatives of 2021–2023, which include master data management, connecting customer journeys, reshaping customer focus, reengineering knowledge, and automating and orchestrating through intelligent automation.

Telecom leaders leverage intelligent automation to pursue multiple objectives in addition to **Efficiency**. **Enablement** gains are exponential, even potentially infinite. **Effectiveness** and **Enablement** gains are closely related to executive leadership recognising IA as a strategic business lever requiring investment, not just a piece of technology.

Telecom leaders demonstrate how intelligent automation can progress the knowledge switch from service operations to customer needs. In particular, intelligent automation is being used as the central engine to overhaul, refashion and enhance the business value of contact centres.

13

Intelligent Automation in Healthcare

The promise of artificial intelligence in medicine is to provide composite, panoramic views of individuals' medical data; to improve decision making; to avoid errors such as misdiagnosis and unnecessary procedures; to help in the ordering and interpretation of appropriate tests; and to recommend treatment.

Eric Topol, '*Deep Medicine:*
How Artificial Intelligence Can Make Healthcare Human Again'

Introduction

During 2020–2022, spurred by the COVID-19 pandemic, intelligent automation galvanised many healthcare organisations: creating new processes, and running operations and services more quickly, cheaply and more effectively, with resilience. But has this been a quick technological fix, or the kick-start of something more lasting? Despite multiple obvious uses for automation, the healthcare sector has largely been a follower, even a laggard. Will increased investments—up 70% in 2023—translate into impressive results or end in automation fatigue?

By 2021, according to Accenture, 69% of healthcare organisations were piloting or had adopted intelligent automation. During 2021, in the USA and UK, a majority of processes being automated were around patient journeys, COVID-19 testing, appointment scheduling and management, patient data extraction and review, claims administration, medical procedure coding

L. P. Willcocks et al., *Maximizing Value with Automation and Digital Transformation*, https://doi.org/10.1007/978-3-031-46569-7_13

and billing, and payment cycle management. HR on- and off-boarding has also been a big use case.

Such positive automation applications during 2020–2022 suggest an obvious forward agenda. First, devising much simpler, more intuitive administrative processes for payments, patient records, and claims. Second, streamlining coordination between diverse care agencies, including insurers, patients, care providers, for preventive and therapeutic care. Third, and most obviously, moving to whole patient care—simplifying and speeding 'moment of truth' service, and building electronic medical records as a foundation for improved management, quality of care and patient outcomes.

But why is automation needed? The big picture tells the story. The World Health Organisation predicts a shortfall of around 9.9 million healthcare professionals worldwide by 2030, despite the global economy creating 40 million new health sector jobs by the same year. Larger, ageing populations, and increasingly complex healthcare demands and therapies, will create rising pressure on relatively fewer health workers. Yet we know that intelligent automation can help. The Brookings Institute estimated that around 33% of tasks currently performed by healthcare practitioners have the potential to be automated.

But healthcare applications can and must go way beyond just improving administration. Intelligent automation, allied with other emerging digital technologies, can transform patient experiences, access and outcomes. It can also transform self-care and prevention, diagnosis and triage, clinical decision support, care delivery and chronic care management. It can also support further healthcare research and innovation.

In fact, they already are. Let us look at the inroads three leading healthcare organisations are making into these possible futures.

Health Care: Possible Futures

Case 1: UK

A new 'digital native' health care provider operates as a GP practice in the UK's National Health Service, as well as a direct care provider in the US serving all economic levels from private practice to Medicaid. Its explicit goal is to disrupt existing systems by providing affordable, accessible, value-based care with a simple, clear strategy: delivering robust clinical outcomes and improved customer experiences while controlling cost. Or, as the company expresses the formula:

Clinical Outcomes + Customer Experience ÷ Cost

The provider first deployed its RPA solutions in UK, where it built core processes, infrastructure and capacity to support 100 k patients. The company developed fully-automated processes for symptom checking (clinical team plus records), digital appointments (including by mobile), and after-care appointment processes. RPA plays a major role in making all these processes more efficient, as well as in clinical operations—coordinating care across patient special needs. It then rolled the automation program to other regions, focusing on core goals. The company reports it is achieving its projected cost and headcount targets in the context of greatly rising volumes, as measured by appointments processed. On basic operating performance, the company's planned **Efficiency** gains were 30% but it reports having achieved 100% improvement—*more than triple expectations.*

Intelligent automation is delivering other kinds of value as well, including the ability to challenge and re-design all 'as-is' processes for greater enterprise **Effectiveness**. The company reports much higher throughput and volumes, with many processes now running 24/7, and much stronger IT focus on core infrastructure processes. Additionally, they have significantly improved compliance performance in HR and Finance functions through automation. Intelligent automation has ultimately enhanced human value, through proactive, easy-to-access care, for example by automating tests and data capture for high-volume services such as COVID-19, diabetes and smears. Productivity and security—for patients and clinical processes—have been much improved through automation, resulting in much higher customer and employee satisfaction. While the company had anticipated **Effectiveness** gains in the region of 50% from their automation program, they report actually getting 200% gains—*four times expectations.*

Beyond **Efficiency** and **Effectiveness**, the company has also realised major benefits in **Enablement.** Intelligent automation has created a platform for innovation, generating analytics for new products/services, and transforming the way healthcare is delivered by giving employees and patients a differentiated experience. Automation has also enabled the company to accelerate time to market for new services and expand market penetration by rapidly implementing innovations and automations across the enterprise. RPA in particular has been instrumental in meeting enforced fast volume ramp-ups for services required to address COVID-19. While the company initially assumed limited **Enablement** gains of around 20%, they report actually getting 200% gains—*10 times expectations.*

Case 2: USA

This US health care network provides end-of-life hospice care, an acute high-intensity service for patients, families, and caregivers at a critical juncture. Serving 10 regional facilities, the company focuses on value-based care, delivering a superior quality, differentiated service with highly trained, dedicated staff. Its experience demonstrates that intelligent automation is not just for large multinationals, but delivers real value across all sizes and sectors.

Two complementary aspirations shape and guide the company's automation strategy. The first is to fully leverage the valuable professional licensure skills of staff in delivering more direct care for patients by freeing them from administrative scheduling and record keeping tasks. Secondly, the provider seeks to differentiate itself on quality of personalised care from private sector companies focused on acquiring non-profit hospice organisations and reducing costs while pricing at previous levels.

The provider chose a cloud-based automation platform for ease of deployment and management, and because it offered access to pre-built technologies and software standardised for health records, reporting and compliance. Business Process Management (BPM) and machine learning tools are being trialled, and enterprise business process modelling software is helping standardise and optimise processes across all branches.

While early on its journey, the provider is already seeing impressive results. On **Efficiency**, this organisation planned for 20% improvement and has got so far 60%, three times expectations, with more to come. The gains have been mainly in security, accuracy in services coding, and avoiding rework, all of which support more robust revenue streams. The company is currently automating parts of processes in ways that allow it to process much more work with the same staff, freeing its highly skilled professionals to do more of what they're trained for—delivering high quality patient care.

Effectiveness is work in progress, with 10% improvements planned and, so far, realised, but more expected. The provider is getting value from wider workforce augmentation, more throughput, and improved regulatory compliance, with a big payoff from much higher employee engagement now freed from routine repetitive tasks.

The organisation had not planned any **Enablement** gains but estimates the actual improvements to date at 20%. The main gains so far have been a fully optimised workforce, enriched employee experiences, higher levels of patient care. Patient experiences have also improved in several areas. For example, automation enables early identification and pre-emption of additional government-approved 'service intensity add-ons' for patients in the last few days of their lives. In the longer term, the company also sees opportunity to provide services to a wider market through a national hospice cooperative. Finally, the company is more competitive, though this is harder to track, because it is one of the few organisations in its sector focusing on intelligent automation thus creating a performance gap.

Case 3: Europe

This healthcare provider initially sought to capture **Efficiency** value from what it called 'automation arbitrage'—using automation to reduce inefficiencies resulting from what it terms 'growth through evolution rather than design.' From 2016, the company focused on automating stable processes and reducing the transaction volumes sent to its BPO provider. One solution involved disaggregating the patient and customer on-boarding process, applying digital workers to handle complex transactional parts of the process, while relying on the BPO provider to handle parts requiring human interpretation. The 24×7 availability from automation was greatly and necessarily strengthened during the pandemic crisis. By 2022, the company had planned for and was getting some 150 percent **Efficiency** gains.

The automation targets have also moved to capture greater enterprise **Effectiveness** by automating other processes, creating what it calls an 'evolving business case.' As we have seen elsewhere, automation greatly improved regulatory compliance through transaction accuracy, traceability and auditability. New automated processes handled new work and work spikes, without increasing headcount. Process improvement occurred as a by-product of automation and external pressures. The company reports that while it has not assigned any value in advance, its gains in **Effectiveness** were 150%—doubling its **Efficiency** gains.

Significant gains from automation also came in the **Enablement** category, in its ability, capability, and resilience to respond to a massive, unanticipated event. The company operated 130 residential care homes—each of them a potential transmission 'hotspot' when the COVID-19 pandemic suddenly hit. Intelligent automation enabled development and deployment of a robust, end-to-end testing regime and response at speed and scale, commensurate with the rapid spread of infections. With tens of thousands of tests required for patients and caregivers every month, the company developed automations to collect, process and triage the tests within seconds, on a 24-h basis, and to send text message alerts instantly to front line managers to isolate identified cases for treatment. Moreover, having collected over three million data points from this activity by March 2021, the company subsequently was gaining a valuable overview of events in real time using analytics.

In practice, **Enablement** gains were unplanned but the company estimates the value gained conservatively at 300% and expect it will probably reach 450% by mid-2021. The head of automation stressed that most often **Enablement** value from automation was not easy to identify initially, but emerged powerfully once the right direction had been established and investments made.

Conclusion

Healthcare organisations are not today unusual in being awash with massive, growing and potentially very valuable data. The COVID-19 experience helped to showcase the importance of accelerating the long haul to digital transformation. The global artificial intelligence in healthcare market size is projected to reach US$187.95 billion by 2030 at a CAGR of 37% across 2022–2030. Will this money be spent wisely? The promise is immense. Probable IA/AI developments are in areas such as: cost effective, errorless treatments, data-driven more personalised healthcare services, possible emotion-AI for mental health disorders and autism, improved drug discovery, smart pills removing invasive procedures, treatment of heart disease, detection of cancer, and diabetes care. More immediately, leaders in our research demonstrate that:

1. Healthcare needs to apply intelligent automation not just to internal administrative processes, but also to interconnectedness between providers, related agencies, government departments and patients.
2. Intelligent automation has a valuable role to play in transforming how patient data is collected, organised and used in order to greatly improve the patient experience.
3. Competitiveness amongst not-for-profit and profit-making healthcare organisations can be a spur to further automation and performance improvements.
4. The data and technology legacies, and the multiple pressures on healthcare organisations make automation very challenging, but ultimately also even more urgent. The 2023–2030 growth in demand for healthcare is projected to be massive, and likewise the pressure on limited resources. As in other sectors, intelligent automation has become a vital coping set of technologies, and for the long run.

14

AI: Ethical and Social Responsibility Challenges

It is time, we think, not just to celebrate and exploit, but also to start issuing serious digital health warnings to accompany these machines.

The Authors

Introduction

Intelligent automation and related technologies—more often referred to in the all-embracing media shorthand as 'AI'—have been the subject of increasing high profile concern. The emergence of ChatGPT, and its very quick adoption by several hundred million users in early 2023, only served to ring more—and even louder—alarm bells. As we have pointed out in earlier chapters, the term 'AI' itself is highly misleading. At present, and at no time in the next decade, will it equate with human intelligence. At best today it refers to the use of machine learning, algorithms, large data sets, and traditional statistical reasoning driven by prodigious computing power and memory. It is in this sense that we will use the term here. The media and businesses subsume a lot of other lesser technologies under the term 'AI'—it's a nice shorthand, and suggests a lot more is going on than is the case. In this book we give multiple examples of impressive uses of these technologies. But it really is important to recognise that impactful technologies like AI are invariably double-edged, and come with consequences that are not necessarily bad, but are often dangerous, and especially when bad actors get involved. In

L. P. Willcocks et al., *Maximizing Value with Automation and Digital Transformation*, https://doi.org/10.1007/978-3-031-46569-7_14

this chapter we want to provide some warnings about the imperfections and limitations of AI.

Chat GPT etc.: Not Born Perfect

ChatGPT is representative of a number of natural language processing tools driven by AI technology. It allows you to have human-like conversations and much more with the chatbot. The language model can answer questions and also assist you with tasks like composing emails, essays, and code. On initial use ChatGPT is impressive. But in the chat function and in the writing essays section we tried it does get things wrong. You could write this off to the fact that it's officially in the 'research' stage. The problem here is that software is never out of the developmental stage, correcting mistakes, and updating to fit with more powerful technologies. That's partly why we have had version 3, 4, 5 … no doubt infinitum. This would not matter except for two facts. Firstly, ChatGPT will have mass adoption—it already has a strong user base, not least because, at least initially, it's free. Secondly, most of the users will be naïve as to what goes into this AI software. They will not understand its limitations.

Stepping back, we need a better steer on what the limitations of AI are. As a counter-weight we look at how these technologies come with in-built serious practical and ethical challenges. To start off, here are three things you need to bear in mind:

1. **AI is not born perfect**—These machines are programmed by humans (and possibly machines) that do not know what their software and algorithms do NOT cover, cannot understand the massive data being used, e.g. biases, accuracy and meaning, do not understand their output, and cannot anticipate the contexts in which the products are used, nor the likely impacts.
2. **AI is not human and never will be**—despite all our attempts to persuade ourselves otherwise. The 'brain like a computer, computer like brain' metaphor is shown by recent neuroscience to be a thin and extremely misleading simplification. And our language tricks us into saying AI is intelligent, feels creates, empathises, thinks, and understands—it does not and will not—because AI is not alive, and does not experience embodied cognition. Meredith Broussard, in her book *Artificial Unintelligence*, summed this up neatly: *"If it's intelligent it's not artificial, if it's artificial, it's not intelligent."*

3. **Ethical and social responsibility challenges.** As today's ChatGPT XE "ChatGPT" examples illustrate, the wider its impact, the more of these challenges AI creates—and they are very large.

An AI Imperfections Test

For this book we have created what we are calling an AI (Artificial Imperfections) Test to guide practical and ethical use of these technologies. There are nine benchmark points in this test—see Fig. 14.1. Let's look at these one by one.

AI is Brittle—meaning AI tends to be good at one or two things rather than, for example, the considerable flexibility/dexterity of humans and their skills. You will hear plenty of examples of what AI cannot do. To quote the Moravec paradox (Moravec 1988):

> it is comparatively easy to make computers exhibit adult level performance on intelligence tests or playing checkers, and difficult or impossible to give them the skills of a one-year-old when it comes to perception and mobility.

ChatGPT is very attractive because it is unusually adept and widely applicable, due to its fundamental natural language model and the vast data lake it has established and continually updates.

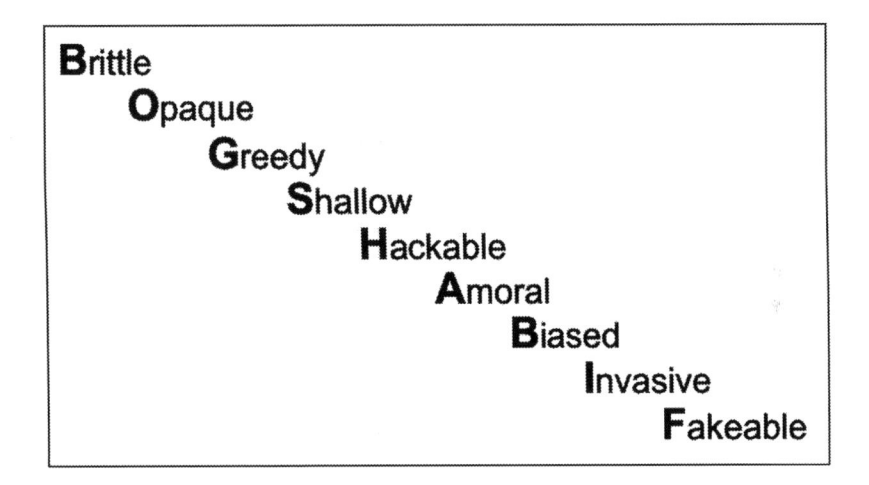

Fig. 14.1 Key technical and ethical 'AI' challenges

AI is Opaque—forming a technology 'black box' it is not clear how AI works, how decisions are made, and how far its judgements and recommendations can be understood and relied upon. A lot of people are working on how to counter the intangibility and lack of transparency that comes with AI. They also have to wrestle with the immense data loads and the speed AI operates at—much of it beyond human understanding. In such a world, small errors—which are not easily identifiable—can accumulate to massive misunderstandings. ChatGPT is an open field for these problems.

AI is Greedy—it requires large data training sets, and thereafter is set up to deal with massive amounts of variable data. Processing power and memory race to keep up. The problem is that a great deal of data is not fit for purpose. And bad data can create misleading algorithms and results. The idea that very big samples solve the problem—what is called Big Data—as used in ChatGPT, for example, is quite a naïve view of the statistics involved. It is not really possible to correct for bad data. And the dirty secret of Big Data is that most data is dirty.

AI is Shallow—it skims the surface of data. As said earlier, it does not understand, feel, empathise, learn, see, create or even learn in any human sense of these terms. Michael Polanyi is credited with what has become known as the Polanyi Paradox—people know more than they can tell. Humans have a lot of tacit knowing that is not easily articulated. With AI there is actually a Reverse Polanyi Paradox: AI tells more than it knows, or rather, more accurately, what it does not know.

AI is Hackable—and eminently so. It does not help that well-funded state organisations are often doing the hacking. Online bad actors abound. The global cybersecurity market continues to soar, reaching some $US 200 billion by 2023 with a 12% compound annual growth rate thereafter. Of course AI can be used both to support but also hack cybersecurity.

AI is Amoral—it has no moral compass, except what the designer encodes in the software. And designers tend not to be specialists in ethics or unanticipated outcomes.

AI is Biased—every day there are further examples of how biased AI can be, including ChatGPT and similar systems. Biases are inherent in the data collected, in the algorithms that process the data, and in the outputs in terms of decisions and recommendations. The maths, quants and complex technology throw a thin veil of seeming objectivity over discriminations that are often misleading (e.g., predictions of future acts of crime) and can be used for good or ill.

AI is Invasive—Shoshana Zuboff leads the charge on AI invasiveness with her recent claim that *"privacy has been extinguished. It is now a zombie."* AI, amongst many other things, is contributing to that outcome.

AI is Fakeable—we have seen plenty of illustrative examples of successful faking. Indeed, with a positive spin, there is a whole industry devoted to this called augmented reality.

Looking across these challenges and AI's likely impacts, you can see that technologies like Chat GPT contain enough practical and ethical dilemmas to fill a text book. New technologies historically tend to have a **duality of impacts**—simultaneously positive and negative, beneficial and dangerous. With emerging technologies like AI and ChatGPT we have to keep asking the question—does 'can' translate into 'should'? In trying to answer this question, Neil Postman suggested many years ago the complex dilemmas new technologies throw up. We always pay a price for technology. The greater the technology the higher the price. There are always winners and losers and the winners always try to persuade the losers that they are winners. There are always epistemological, political and social prejudices imbedded in great technologies. Great technologies (like AI) are not additive but ecological— they can change everything. And technology can become perceived as part of the natural order of things, and therefore controls more of our lives than may be good for us. These wise warnings were given in 1998, and have even more urgency today.

Consider ChatGPT and other kinds of generative AI based on large language models (LLMs). Their likely massive impacts make only worse that they can go wrong in a multitude of ways. Even in 2023, generative AI was adding to the sort of harms perpetrated on the internet every day. Given their relatively open access, there is no reason why they cannot be used for less tha beneficial purposes. If they single mindedly pursue a specific goal set by a user, they might pursue that goal in ways and with outcomes not originally desired.

A lot of effort will have to go into producing seemingly safe models, but the systems themselves, and bad actors might well find ways round safety measures like having a secondary AI watching over output, reinforcement learning from human feedback, and vetting the models more thoroughly before released. Then there is the sheer cost in terms of computing power, labour, data sets and electricity. Training language learning models get more expensive faster than the models get better (GTP-4 cost around $US 100 million). Moreover, how much more high quality language data will become available? For example, the likelihood of intellectual copyright infringements have arisen already. Will this lead to a lowering of standards for admissible data? A more profound question needs to be asked where even by 2023 the intended and emergent capabilities of the biggest models would seem to have outpaced the designers' understanding and control. On the evidence so far, one wonders whether there can ever be a 'safe' generative AI.

Conclusion

Our conclusion and fear are that collective responses to AI display an ethical casualness and lack of social responsibility that put us all in peril. Once again professional, social, legal and institutional controls lag almost a decade behind where accelerating technologies are taking us. It is time, we think, not just to celebrate and exploit, but also to start issuing serious digital health warnings to accompany these machines.

Part III

Digital Transformation

15

Managing the Digital Catch-22

When digital transformation is done right, it's like a caterpillar turning into a butterfly, but when done wrong, all you get is a really fast caterpillar.

George Westerman

Today's organisations are facing a digital catch-22. On the one hand, digital transformation is difficult and costly, and short-term investment may be needed elsewhere to where it's really hurting. On the other hand, today's organisations cannot afford not to become tomorrow's digital businesses. In this chapter we will point up the dimensions and intractability of this digital catch-22, before suggesting some ways forward.

But not so fast; firstly, what is **digital transformation**? Digital transformation requires **digitisation**—converting something non-digital (e.g., a health record, an identity card) into a digital format that can then be used by a computer system. Digital transformation also requires **digitalisation**—enabling, improving, or changing business operations, functions, or activities by utilising digital technologies and using digitised data to create management intelligence and actionable knowledge. All three—digitisation, digitalisation, and digital transformation—are needed to build a **digital business**. Digitisation and digitalisation are necessary but insufficient.

Digital transformation must focus on the whole organisation, and large-scale change. It involves radical redesign, then deployment of business

L. P. Willcocks et al., *Maximizing Value with Automation and Digital Transformation*, https://doi.org/10.1007/978-3-031-46569-7_15

models, activities, processes, and competencies to fully leverage the opportunities provided by digital technologies. We would guess you already have some idea of why it is so difficult.

Let's find some more evidence for the high level of challenge (see also Chapter 1). It is notable, firstly, that organisations are surprisingly slow into digital transformation, given that this has been on many executive agendas since at least 2010. Many organisations digitise, digitalise even, but this does not add up to digital transformation, though many might think it does. The reasons for the lack of speed are complex, but failure is four times more likely than success. The high failure rate is indicative of the large number of stumbling blocks and can be very dissuading for others. A point we have made before: slow progress reflects how 'siloed' many organisations have become. What we call the 'eight-siloed organisation' points to the barriers to change inherited from older business models. As discussed in Chapter 1, the siloes include processes, technology, data, culture, structures, strategy, skills sets, and managerial mindsets. When it comes to digital transformation, any organisation with all these siloes is severely hamstrung from the start.

Furthermore, very few organisations have all the automation execution and digital transformation capabilities needed for success. We detail these in Chapter 17. As we will suggest in Chapters 19, building digital platforms is vital but evolutionary and challenging. Many organisations also have a 'blind spot' in that their automation efforts do not connect up and integrate with their digital transformation agendas and activities (see Chapter 18). Finally, change management, culture and the politics of technological change are invariably difficult to negotiate and get right. These are, therefore, the subjects of Chapters 21.

Yet there is another side to the digital catch-22. What happens if, putting it colloquially, you fail to sail? There are relatively few best performers on digital transformation. As discussed in Chapter 1, these are getting disproportionate gains, recording markedly higher profitability and revenues, accelerating away from the others, and may well establish a competitive advantage that becomes irreversible. What are they achieving? According to one study, they had increased the agility of their digital-strategy practices, enabling first-mover opportunities. They had taken advantage of digital platforms to access broader eco-systems and innovate new digital products and business models. They had utilised mergers and acquisitions to build new digital capabilities and digital businesses. A significant feature was that they had invested ahead of their competitors in digital talent.

Our own work suggests that the best performers on digital transformation add up to around only 20 percent of organisations, all recording up to a

30 percent increase in revenues as a part outcome of their digital technology investments over the previous four years. They come from most sectors and regions of the world and are not limited to the obvious hi-tech US and Chinese firms. To add even more urgency, our evidence, consistent with other studies, shows that being slow to adopt digital technologies may reduce risk in the short term, in terms of cost, talent and time, but builds growing business risk and reduced competitiveness in the long term. And this trend will be repeated and magnified during the growing adoption of automation and 'AI' over the next seven years.

Reporting in 2023, an Accenture study showed the gap between top and bottom digital performers ('leaders' and 'laggards') had grown to 5× revenue growth (from 2× in 2019), with leaders doubling investments faster than before. Interestingly they also identified a small new group—called 'leapfroggers'—who compressed their digital transformations to convert the pandemic's challenges into new opportunities. 'Leaders' and 'leapfroggers' shared three characteristics—moving to the cloud and embracing new SMAC/BRAIDA technologies (see Chapter 16); shifting their IT budgets to innovation over maintenance; and focusing on creating broader value (for example, touching twice as many processes and focusing on areas like training).

So, yes evidence accumulates for a digital catch-22. But was an unlikely saviour revealed in the form of the pandemic and economic crisis? These have undoubtedly accelerated corporate moves toward digitisation and digitalisation—primarily to survive in the short term, establish resilience, and to maintain competitiveness. But we found motives mixed, capabilities variable, and planning horizons mostly short term. That said, a McKinsey survey suggests that COVID-19 pushed companies over a technology tipping point. Between January and October 2020, the digitisation of customer and supply-chain interactions and of internal operations had accelerated by three to four years. The share of digital or digitally enabled products in corporate portfolios had accelerated by seven years. Nearly all respondents had put in place quickly at least temporary solutions, to meet many of the new demands on them. Funding for digital initiatives increased more than for anything else.

Moreover, the largest shifts in the crisis were also the ones most likely to stick—think changing customer needs and expectations; more remote working/collaboration; cloud migration; customer demand for online products and services; and increasing spend on security. Those who had invested heavily into digital technologies over the previous three years also reported a range of facilitating technology-related capabilities that others lacked in the crisis. This meant they were better prepared for the crisis.

Did COVID-19, then, make digital transformation easier? Well, the evidence is that the digital catch-22 has not gone away. Digital technologies are gaining a higher profile amongst the executives who make the key decisions, but the difficulties and complexities of large-scale organisational change on many fronts are not easily circumvented (see Chapter 21), and there remain many other pressing matters to deal with, distracting executive attention. Moreover, it is one thing to have a burning platform driving needed organisational change, but the particular emergency conditions during 2020–2022 may not be repeated. The spectre of fading competitiveness may not be sufficient motivation for all too many organisations. Clayton Christensen's 'innovators' dilemma'—"*a successful company with established products will get pushed aside, unless managers know how and when to abandon traditional business practices*"—could kick in once again.

There also remains the management question. For many years, in study after study, we have found that when it comes to introducing information, communication, and now digital technologies into organisations, the majority of challenges and issues—up to 75 percent—are managerial and organisational, not technological. For digital transformation, the major challenge we have identified is getting to understand and develop competencies for large-scale radical organisational change shaped by disruptive technologies. Reporting in 2020, a Michael Wade and Jialu Shan study is consistent with this in pointing to past transformation failures arising from unrealistic expectations, limited scope, poor governance, and underestimating cultural barriers. They found that, in success stories like Cisco, Unilever, DBS bank (see Chapter 22), ABB, Nike and Disney, they all had precise, clear, unambiguous, realistic, succinct, measurable objectives that included everyone in the company.

Other studies add to this. From their findings as at 2022, Gerald Kane and colleagues identify the need for a whole organisation approach, but also stress the neglected role of people in digital transformation. They point to the importance of transformative leadership, developing a digital talent mindset, and making the organisation a digital talent magnet.

It is a useful reminder that no technology is a silver bullet, a fire-and-forget missile or plug-and-play—despite the widespread use of these misleading metaphors. Jeanne Ross, Cynthia Beath and Martin Mocker, reporting in 2021, recognised that for an established organisation, existing organisational structures, legacy systems, and embedded habits are significant obstacles. They suggest an evolutionary approach of gradual componentisation of parts of the business, producing digitised business operations and units that fit together over time, building towards creating a digital business. They offer the example of Carmax, a US$20 billion used car retailer and wholesaler that benefited between 2015 and 2020 from gradual, but radical redesign of its systems, processes and people. Meanwhile, Jacques Bughin and colleagues (2020) suggest several practices for organisations in catch-up mode (see also Chapter 19): make your strategy process more dynamic; lay out clear priorities; invest early and aggressively in requisite capabilities and talent (especially at senior executive level); invest the time and money—this will only happen if becoming digital is a top priority amongst corporate leaders; redefine how you measure success; and empower people. Digital transformations were more likely to be exceptionally effective when companies gave people clear roles and responsibilities and put an 'owner' in charge of each transformation initiative.

All this suggests that there are ways out of the digital catch-22, but senior executives responsible for digital transformation will need to take a much bigger view of the change process, if the potential business value of digital investments is going to be realised. All this was further complicated by the challenging 2023 business climate globally, fuelled not least by geo-political tensions. Faced with market downturns, possible recession and counter globalisation trends, many businesses understandably headed for cover. As we noted in Chapter 1, 'digital transformation' became de facto cost transformation. However, a minority of organisations also took another view and sought limited, but more expansive goals. Digital transformation typically seeks radical changes in customer experience, operational efficiency and the creation of new business models, products and services and new markets. In the 2023 environment, financial managers became much more focused on customer retention, capital strength and the viability of the cost structure in the downturn. It became more realistic to pursue one or other of these goals—usually customer experience and/or operational efficiencies, using digital technologies. This diluted but did not eliminate the digital catch-22.

Conclusion

In summary, there are several interesting features in the long trend of moving to digital business. By 2015 most large organisations were espousing digital business strategies. But by 2021, if you exclude the 'born digital' and the obvious hi-tech companies, few had become digital businesses. According to an Accenture 2022 study, over 80 percent of organisations had tended to focus on transforming parts of their business rather than the whole, and tended to treat transformation as a finite program rather than a continuous process. In the 2023 global business environment, this perhaps had become a more viable temporary option. But if so, these 'follower' organisations will have to accept less startling results. In comparison, by 2023 digital leaders were already getting 10 percent higher incremental revenue growth, 13 percent higher cost-reduction improvements and 17 percent higher balance-sheet improvements. They also gained 1.3 times more financial value in the first six months, and were delivering their digital transformation strategy quicker. Consistent with our Chapter 1 findings, the Accenture study found they also performed a third better on sustainability, customer, supplier and employee experiences, and higher on innovation, outcomes for talent, and inclusion and diversity.

On our evidence, for the sizeable majority of organisations, digital transformation will still be serious work-in-progress, even beyond 2025. This highlights first that organisational change can take a very long time. Secondly, it underlines the importance of execution capabilities, without which strategy means very little. The third point of note is how an ostensible support activity—IT —can move to the core of the business and become not just a strategic weapon, in terms of cost efficiency and differentiation, but a fundamental platform for operating. The fourth observation is that the 2020–2023 pandemic and economic crises accelerated the adoption of digital technologies and the moves to digital business but left much work to be done if organisations are to enhance competitiveness, streamline operations, improve organisational resilience, and become digital businesses.

16

On the Yellow Brick Road—The Ten Technologies that Pave the Way

The Web as I envisaged it, we have not seen it yet. The future is still so much bigger than the past.

Tim Berners-Lee, inventor of the World Wide Web

We are analogue beings living in a digital world, facing a quantum future.

Neil Turok

Introduction

The general view is that the 2020–2022 pandemic accelerated the adoption of automation, and moves towards digital business. The case for these technologies was well made by early 2020s experiences. The technology worked. It provided alternative ways of working in a crisis. It allowed a high degree of virtuality, and all the upsides to that, in a world rendered physically semi-paralysed, albeit temporarily. It would be a foundation for building resilience in the face of increasingly uncertain business environments, and future crises. But when we talk of the technology, in fact there are many emerging digital technologies. Which to bet on? Our own view is that ten technologies, which we call SMAC/BRAIDA, were the most obvious candidates as at 2023–2024, and that, used combinatorially, they represent the basis for most businesses in advanced economies being digital by 2030. That long? we hear you ask.

L. P. Willcocks et al., *Maximizing Value with Automation and Digital Transformation*, https://doi.org/10.1007/978-3-031-46569-7_16

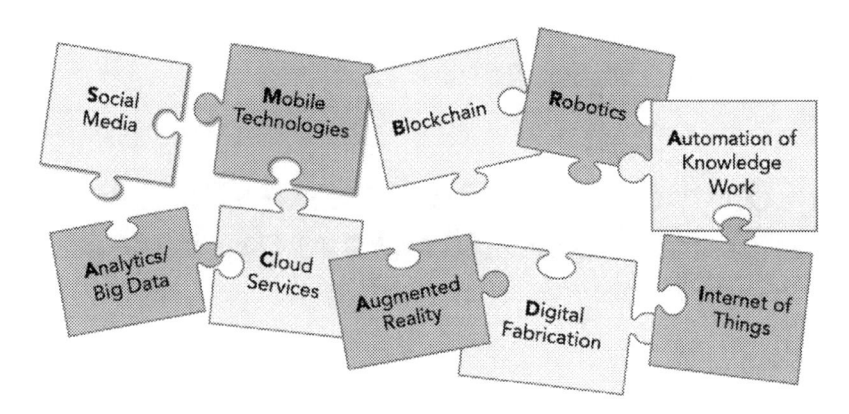

Fig. 16.1 SMAC/BRAIDA Technologies 2020–2030

Well, yes, because digital transformation is highly challenging as we have seen, though failure and abandonment of projects had been reducing somewhat over the 2019–2023 period, and 'failure' may be too strong a word where it also covers not meeting senior executives' business expectations. Moreover, just bringing one of these technologies into an organisation is highly problematic, as has been found in our KCP work on automation. With digital transformation organisations are trying to harness, eventually, all ten, with a watching brief on several more, for example quantum computing, metaverse, Web3, 5G, and bio-engineering. Let's understand their potential and the challenges. Our rough guide begins with Fig. 16.1. Then we take a much wider look at how these connect up in likely major tech trends across this decade.

SMAC/BRAIDA Technologies

Let's look at Fig. 16.1 in more detail:

Social media are interactive computer-mediated technologies that facilitate the creation or sharing of information, ideas, career interests and other forms of expression via virtual communities and networks. Social media are interactive Web 2.0 (moving to Web 3.0) Internet-based applications. User-generated content, or user-shared content, such as text posts or comments, digital photos or videos, and data generated through all online interactions, is the lifeblood of social media. Users create service-specific profiles and identities for the website or app that are designed and maintained by the social media organisation. Users usually access social media services via web-based technologies on desktops and laptops, or download services that offer social media functionality to their mobile devices (e.g., smartphones and tablets. As users engage with these electronic services, they create highly interactive platforms through which individuals, communities, and organisations can share, co-create, discuss, participate and modify user-generated content or self-curated content posted online.

Social media has a massive and ever-increasing business and user base so the application of technologies will accelerate across this decade. Expect more connectivity; more user-friendly technologies; dynamic, changing usage trends; more platforms and services; and more smartphone and tablet computers applications. Businesses will massively expand their social media technologies as new tools appear for marketing, communication, sales, promotions and discounts, loyalty programs and e-commerce, whilst also using it internally, for communication and employee learning, for example. Organisations also increasingly use social media to monitor, track, and analyse online conversations about their brand or products, or about related topics of interest—a trend with seemingly limitless options and uses, for sentiment tracking, industry trends, market predictions, and tracking insider trading as just some examples. Despite many likely run-ins with regulation, the massive growth in social media use during this decade seems inescapable.

Mobile technology is used for cellular communication. Mobile technology has evolved very rapidly over the past few years. Since the start of this millennium, a standard mobile device has gone from being no more than a simple two-way pager to being a mobile phone; GPS navigation device; an embedded web browser; an instant messaging client; and a handheld gaming console. Many experts believe that the future of computer technology rests in mobile computing with wireless networking. Mobile computing by way of tablet computers is becoming more popular. Tablets are available on the 3G, 4G and 5G networks. In 2022 the estimated number of smartphone users worldwide was 6.64 billion with 88 percent of mobile time spent in applications. Futurists share one prediction—mobile everywhere—and they are probably right.

The compelling trends are towards ever more apps and greater connectivity with other digital technologies. For example, with AI there are already chatbots and virtual personal assistants in service industries, personalised ads and recommendations in eCommerce, financial forecasting solutions, voice recognition apps for hands-free communication, gaming and entertainment, and motion and facial detection apps for surveillance systems. Expect a continuing trend towards mobile apps that recognise voice commands; analyse textual and visual data; anticipate user behaviour; and make forecasts, recommendations, and decisions. Mobile is also linking with augmented and virtual reality, moving from gaming apps, into practical areas such as navigation, interior design, education, object measurement—creating a likely US$200 billion market by 2030. Also expect more mobile-Internet of Things (see below) connectivity for both business to consumer and business to business applications.

Mobile payment is another application, with two billion users in 2021—accelerated by the 2020–2022 partial pandemic shutdowns globally—and much wider adoption by customers and businesses since. Mobile apps will also become much more cloud based where it provides more storage, computing power, security, cheapness, and increasing customer retention, streamlining of operations.

5G is also transforming the world of mobile app development, offering connectivity, bandwidth, low latency (allows real time data processing) and speed. We are also seeing wearable technology impacting on mobile development. All this requires further developments in cross-platform enabled applications, and in mobile security.

Analytics concerns the discovery, interpretation, and communication of meaningful patterns in data. It also entails applying data patterns towards effective decision-making. Especially valuable in areas rich with recorded information, analytics relies on the simultaneous application of statistics, computer programming and operations research to quantify performance. Organisations apply analytics to business data to describe, predict, and improve business performance. Specifically, areas within analytics include predictive analytics; prescriptive analytics; enterprise decision management; descriptive analytics; cognitive analytics; Big Data analytics; retail analytics; supply chain analytics; store assortment and stock-keeping unit optimisation; marketing optimisation and marketing mix modelling; web analytics; call analytics; speech analytics; sales force sizing and optimisation; price and promotion modelling; predictive science; credit risk analysis; and fraud analytics. The algorithms and software used for analytics harness the most current methods in computer science, statistics, and mathematics.

AI is set to become a fundamental driver of business analytics, but needs to evolve into a more responsible and scalable set of technologies as organisations will require a lot more from AI-based systems. At its best AI can offer real time, online, scaled, pro-active analytics capabilities. AI can also facilitate more plain language reporting, as can data discovery using visuals. Faced with the exponential data explosion expect developments in these areas. Inevitably there will more advances in and adoption of predictive and prescriptive analytical tools. Expect more collaborative business intelligence, more uses of natural language processing, greater data automation; also, more embedded analytics—where data analytics occurs within a user's natural workflow (expected to be a US$80 billion market in 2026). All this is going to have to come with much greater investments in data security and governance.

Cloud computing is a foundational technology for the others. This is the on-demand availability of computer system resources, especially data storage and computing power, without direct management by the user. The term is generally used to describe data centres made available to many users over the Internet. Cloud computing is usually bought on a pay-per-use basis from a cloud vendor, for example Amazon's data services. Today's large clouds often have functions distributed over multiple locations from central servers. Clouds may be limited to a single organisation (enterprise clouds) or be available to many organisations (public cloud). Cloud computing relies on sharing of resources to achieve coherence and economies of scale. The advantages of cloud computing are said to be: companies avoid or minimise up-front IT infrastructure costs; it allows enterprises to get their applications up and running faster; improved manageability and less maintenance; and it enables IT teams to more rapidly adjust resources to meet fluctuating and unpredictable demand. On the downside, people cite perceived insecurity (especially with the public cloud), regulation compliance problems, difficulties integrating with existing IT and business processes, in-house skills shortages, contracting and pricing being sometimes inflexible, and lock-in to suppliers.

The cloud services market amounted to US$220 billion in 2022 with growth slowing in 2023, but anticipated to rise with exponential data growth to 2030. Also becoming cloud first is critical to digital transformation, offering customers with new digital experiences that are faster to market, while reducing costs, providing analytics, simplifying innovation and scalability and, if managed well, reducing risk. According to one estimate, cloud computing could realise, for Fortune 500 companies, more than US$1 trillion in value by 2030—and not just from cost savings. High-tech, retail, and healthcare organisations occupy the top end of value capture but the technologies are becoming sufficiently adaptive for use in most sectors, though heavily regulated industries (for example finance, utilities and insurance), understandably show a deal of wariness.

Blockchain has been surrounded by much excitement, mainly because of the high-profile debates about crypto-currencies. A blockchain consists of a growing list of records, called blocks, which are linked using cryptography.

Each block contains a cryptographic hash of the previous block, a timestamp, and transaction data. By design, a blockchain is resistant to modification of the data. It has been described as an open, distributed ledger that can record transactions between two parties efficiently and in a verifiable and permanent way. For use as a distributed ledger, a blockchain is typically managed by a peer-to-peer network collectively adhering to a protocol for inter-node communication and validating new blocks. Blockchain technology has already been used as a basis for many cryptocurrencies, and for smart contracts, in supply chains and in financial services, for example distributed ledgers in banking, and in banking settlement systems. Proponents suggest it has massive business potential, but also that will take time to have large-scale impacts.

Indeed, apart from cryptocurrency applications, blockchain is a slow train—but it is coming. The global blockchain market was valued at US\$4.67 billion in 2021 and is projected to grow from US\$7.18 billion in 2022 to US\$163.83 billion by 2029, exhibiting a compound annual growth rate (CAGR) of 56.3 percent during the forecast period.

The main advantages of blockchain technology are: a decentralised network; a transparent and universal recording system; a decentralised trust chain; no single point of failure; low operational cost; better accessibility; unalterable and indestructible technology; enhanced security; and confidentiality. In turn, the main disadvantages relate to high energy dependence, the difficult process of integration, lesser scalability and the high costs of implementation, together with transactions being slower than with other payment methods. For today, it probably best fits businesses that want to take advantage of the distributed ledger features, but work-in-progress technologies may well make the challenges much less, and improve advantages and adoption over the next few years.

Robotics is the next technology on our list. This is an interdisciplinary branch of engineering and science that includes mechanical engineering, electronic engineering, information engineering, and computer science. Robotics deals with the design, construction, operation, and use of robots, as well as computer systems for their control, sensory feedback, and information processing. These technologies are used to develop machines that can substitute for humans and replicate human actions. Robots can be used in many situations and for lots of purposes, including in dangerous environments, manufacturing processes, and increasingly in service settings, e.g., healthcare. The global industrial robot market was US\$52.34 billion in 2023 and was expected to grow to US\$77.23 billion in 2027 at a CAGR of 10–11 percent. The service robotics market size is projected to reach US\$84.8 billion by 2028 from US$41.5 billion in 2023; it is expected to grow at a CAGR of 15.4 percent from 2023 to 2028. Service robots for personal, professional and business use is a growing area and the uses of robots in service functions are potentially vast, for example: cleaning robots for public places; delivery robots in offices or hospitals; fire-fighting robots; rehabilitation robots; elderly care service robots; and surgery robots in hospitals.

Automation of knowledge work has been KCP's primary research ground. As we saw in 'Introduction', this involves software called robotic process automation (RPA), which automates simple repetitive processes by taking structured data and applying prior configured rules to produce a deterministic outcome, i.e., a correct answer. It is generically applicable to most business processes. Its limitations mean that it is increasingly complemented and augmented by cognitive or intelligent automation software (see 'SMAC/BRAIDA Technologies') that can process unstructured and structured data, through algorithms (pre-built from training data) to arrive at probabilistic outcomes (see Fig. 16.2). For example, in one development Zurich insurance used cognitive technology to check accident claims by looking at medical records supplied by the claimant and pictures of injuries to establish the claim and how much to pay out. The process used to take a human some 59 minutes, but subsequently was automated to take less than 6 seconds—saving dramatically on cost, time while improving accuracy. To reiterate Chapter 1, artificial intelligence (AI) is an umbrella term used to cover all automation technologies, though the vast majority in use as at 2023 did not match the strong definition of AI as 'getting computers to do the sorts of things human minds can do'.

Reflecting on Fig. 16.2, it is important to point out that not all 'RPA'-badged tools are the same. Many suppliers initially provided desktop RPA, where software replicated and automated some task a person carried out on

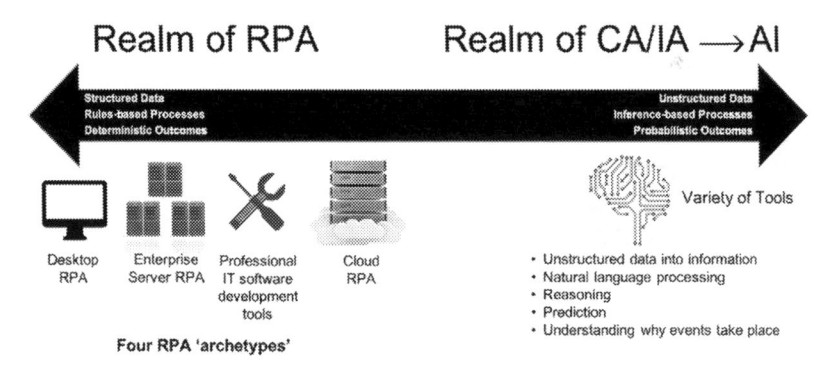

Fig. 16.2 The RPA, cognitive/intelligent automation and AI landscape

the desktop. This increased productivity at the desktop level, but enterprise RPA providers went a step further in adding in software that made the RPA component compatible and safe to use with other technologies and at the enterprise level. Other suppliers provided a skeleton RPA framework which need much more customised software development to make it consistent with an organisation's systems and technology. More recently, some providers have offered RPA services over the cloud, to rent as needed. All suppliers have now embarked on making their software more enterprise compatible, and more cloud friendly, as well as making RPA more usable with cognitive automation tools (see Fig. 16.2 and 'SMAC/BRAIDA Technologies').

These technologies are definitely fundamental for building digital platforms and for digital business. According to some estimates the RPA global market was around US$3.4 billion in 2023 growing to US$31 billion by 2030 registering a CAGR of 38.2 percent. The global intelligent process automation market size was valued at US$9.52 billion in 2021 and has been projected to reach US$37.63 billion by 2030 at a CAGR of 16.50 percent. These figures have to be subsumed into our Chapter 1 figures for the overall global artificial intelligence market size of US$136.55 billion in 2022 rising to US$1.812 billion by 2030, expanding at a CAGR of 37.3 percent from 2023 to 2030.

The Internet of Things (IoT) is another foundational technology, because it collects data. IoT represents a system of interrelated computing devices, mechanical and digital machines (including sensors), provided with unique identifiers (UIDs) and the ability to transfer data over a network without requiring human-to-human or human-to-computer interaction. The Internet of Things has evolved due to the convergence of multiple technologies, real-time analytics, machine learning, commodity sensors, and embedded systems. Real world examples abound. In the consumer market, IoT is used in so-called 'smart homes', but has multiple uses for collecting data through sensors (for example traffic and people monitoring), creating customer intelligence, and building data for business analytics in real time. There are a number of serious concerns about dangers in the growth of IoT, especially in the areas of privacy and security, and consequently industry and governmental moves have been addressing these.

That said, on some estimates the potential value of IoT (as opposed to expenditure on IoT) could amount to some US$12 trillion globally by 2030—very much larger than for many other digital technologies. Making day-to-day management of assets and people more efficient could account for 41 percent of that value, health care uses some 15 percent, human productivity another 15 percent, condition-based maintenance a further 12 percent.

Other value clusters include energy management, sales promotion, safety and security, and autonomous vehicles. More broadly, IoT technology could function as a global infrastructure, enabling advanced services to interconnect things based on existing and evolving communications technologies. It also offers interoperable information and the ability to operate independently without human intervention. As a result, the technology is expected to open up new revenue streams, drive business efficiencies, improve delivery of existing services and enable new business models. Not surprisingly there will be heavy expenditure here. The global IoT market is projected to grow from US$662 billion in 2023 to US$3,353 billion by 2030, at a CAGR of 26.1 percent.

Digital fabrication and modelling are where we move to next. This is a design and production process that combines 3D modelling, or computing-aided design (CAD), with additive and subtractive manufacturing. Additive manufacturing is also known as 3D printing, while subtractive manufacturing may also be referred to as machining. Many other technologies can be exploited to physically produce the designed objects. Digital manufacturing is an integrated approach to manufacture or construct products using a computer system. It uses computer aided design (CAD), computer aided modelling (CAM), Internet of Things (IoT), and Big Data analytics concepts. Digital manufacturing has features such as maximised productivity, improved quality, reduced operating costs, and limited repetitive work. The global digital manufacturing market is expected to reach US$1370.3 billion by 2030, from US$276.5 billion in 2020, registering a CAGR of 16.5 percent from 2021 to 2030.

Augmented Reality (AR) is the final technology to consider**.** Augmented reality is an interactive experience of a real-world environment where the objects that reside in the real world are enhanced by computer-generated perceptual information, sometimes across multiple sensory modalities. The overlaid sensory information can be constructive (i.e., additive to the natural environment), or destructive (i.e. masking of the natural environment). This experience is seamlessly interwoven with the physical world such that it is perceived as an immersive aspect of the real environment. The primary value of augmented reality is the manner in which components of the digital world blend into a person's perception of the real world, not as a simple display of data, but through the integration of immersive sensations, which are perceived as natural parts of an environment. The global AR market was estimated at US$25.33 billion in 2021 and is expected to expand at a CAGR of 40.9 percent from 2022 to 2030. Potential innovations abound that will provide interactive experience to end-users. The proliferation of handheld devices like smartphones and smart glasses as well as business adoption and increase use of mobile AR technology for providing more immersive experiences—these trends are expected to contribute to the growth of the market.

Watching Briefs

At this stage we suggest a watching brief for developments in quantum computing, metaverse, 5G, Web3, 5G, and bio-engineering. As a page holder, **quantum computing** is an emerging field that applies some basic principles of quantum mechanics to process information at radical speeds. Quantum computing involves qubits. Unlike a normal computer bit, which can be either 0 or 1, a qubit can exist in a multidimensional state. The power of quantum computers grows exponentially with more qubits. Classical computers that add more bits can increase power only linearly. Quantum computing requires less power and can execute any task very fast and very accurately compared to a classical computer. However, quantum computing is still at an early developmental stage.

Likewise, the **metaverse**, a network of 3D virtual worlds focused on social and economic connection. Have we not got these already? Well, yes for example virtual worlds and on-line games. But the aspiration with the more recently touted metaverse is much bigger and fraught with a lot more problems, both technological and also social economic and legal. Therefore, we give it a 'watching' brief.

5G refers to 5th generation mobile network. It is a new global wireless standard after 1G, 2G, 3G, and 4G networks. 5G enables a new kind of network that is designed to connect virtually everyone and everything together including machines, objects, and devices. As of 2023, 5G faced a number of challenges including frequency band and spectrum availability issues, deployment issues, upgrading of mobile devices by end users, managing expenses of deployment, and dealing with security and privacy issues.

Web 3.0 is conceptually a new iteration of the World Wide Web incorporating decentralisation, blockchain technologies, and token based economics. Proponents contrast it with Web 2.0 where data and content are said to be centralised in a small group of 'Big Tech' companies. Some suggest Web 3.0 will give increased data security, scalability, and privacy and combat 'Big Tech' power, while others see decentralisation as resulting in dangers to privacy, and lower control over and the proliferation of negative content, while rechannelling power into the hands of a few investors and shareholders. Again, a watching brief.

Bioengineering is where engineering meets medicine by preventing invasive surgery, innovating hi-tech health devices, and enhancing artificial organs. It spans wearable devices, tissue engineering and biomechanics, with career paths to healthcare, pharmaceuticals and scientific research. Bioengineering looks to be a massive potential growth area, though limited to mainly the health and related sectors.

Major Trends

There are multiple trends that cross these technologies. We will restrict the discussion to what we see as the top five needing attention going forward.

One common thread through this chapter is **advanced connectivity**. A 2022 McKinsey study characterised this as based on developments in optical fibre; low power wide area networks; next generation Wi-Fi; next generation 5G/6G cellular protocols and technologies, and low earth-orbit satellite constellations and portrayed these as major value creators in 12 major industries. They reckoned that use cases in just four of these sectors could boost global GDP by $US1.2 trillion to $US2 trillion by 2030. However, we would add that what is difficult to quantify is the potentially exponential value to be gained as the ten major technologies we highlighted above increasingly work in combination.

A second major cross trend is the growing demand for **cybersecurity**. On one estimate amongst many, global spending on cybersecurity products will reach US$1.75 trillion in the next 5 years. During 2020–2022 businesses were forced to shift from physical to digital and relied on online services leaving a gap for cyberattacks. But more broadly the challenges have become much more threatening. A KPMG survey found 73 percent of CEOs saying that geopolitical uncertainty during 2022 was increasing worries over corporate attacks, compared to 61 percent in 2021. However, 24 percent admitted to not being prepared for a potential attack, a figure that increased compared to 13 percent in the previous year. Expect a rise in cybersecurity threats, and a concomitant increase in business, and government, counter-measures.

Thirdly, amongst the major reports, **applied AI** is slated to impact massively—if not everywhere then, at least according to a McKinsey 2022 report, in 17 major economic sectors; we have already indicated above some potential developments. Industrialising AI would definitely accelerate usage.

Inhibiting factors may be ethical, social responsibility and regulatory challenges (see Chapter 14); also, talent and funding shortages and cybersecurity risks.

Fourthly, **cloud and edge computing** developments look to be massive enablers. Expect bigger cloud platforms and massively extended cloud services, based on hugely scaled data storage and computing centres connected by fast, high-capacity networks. Increasingly, these platforms also incorporate the advantages of computational and data resources at network edge nodes located near end users or in their facilities. The technologies in play will be data centres, edge technologies, Internet of Things, and network infrastructure.

A fifth cross-sector trend, related also to the cybersecurity rationale mentioned above, needs to be in the development of **trust architectures and digital identities.** These technologies help organisations to manage technology and data risks, secure assets and accelerate innovation, thus enhancing organisational performance and improving customer relationships. The underlying technologies include zero-trust architectures, digital-identity systems, and privacy engineering. Other technologies also need to build trust, for example Chapter 14 pointed to AI models needing to be more secure, explainable and relatively free from bias.

Conclusion

Looking across these impressive technologies, you will quickly get a sense of their immense potential for being applied in organisations and driving business value, especially when they begin to be used, as some are already, in combination. McKinsey Global Institute has estimated that applying these technologies could add an additional US$13 trillion to global GDP by 2030. But historically there is a long challenging road to implementation of a major technology, let alone its exploitation. On past experience it can take from eight to 26 years for one such technology to be deployed by 90 percent of organisations across sectors. And certainly, that is how it has been playing out so far with cloud computing; robotics; automation of knowledge work; Internet of Things; analytics; with, of course, social media and mobile being adopted more quickly, and blockchain, digital fabrication and augmented reality being more slowly deployed.

17

The Blind Spot

At best, people are open to scrutinizing themselves and considering their blind spots; at worst, they become defensive and angry.
Sheryl Sandberg

Silo thinking creates blind spots and enlarges IT-business gaps.
Pearl Zhu

Introduction

We all have a blind spot. Part of the optic disc lacks some cells to detect light, and to that extent our vision is impaired. Fortunately, based on surrounding detail and information from the other eye, our brain interpolates the missing data, so it is not normally perceived. But is there a similar process in corporations?

While we have researched multiple sectors and found similar findings, here the focus is just on global banking and financial services. Mapping to Fig. 17.1. We have followed a large number of organisations from the 'Beginner' stage. The successful ones manage to scale within and across functions building an embedded enterprise automation capability and gaining ROI up to 200 percent—in rare cases even more. Payoffs are directly related to the strength of the automation execution capabilities applied (we will detail these in Chapter 18). But though task automation by itself helps organisations become faster and smarter, it can only take you so far. At some

L. P. Willcocks et al., *Maximizing Value with Automation and Digital Transformation*,
https://doi.org/10.1007/978-3-031-46569-7_17

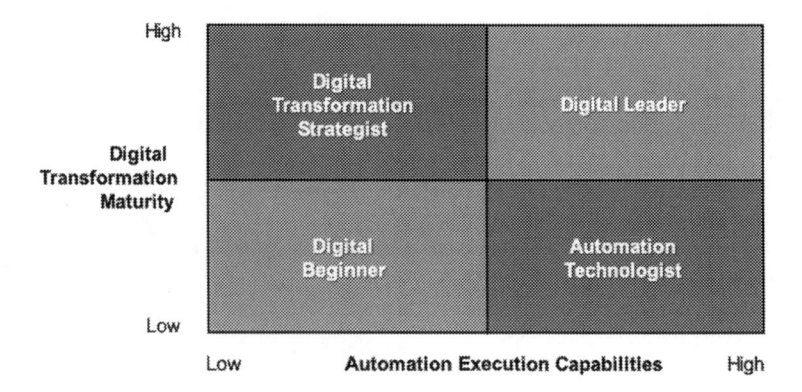

Fig. 17.1 Digital transformation and automation execution maturity (*Source* Authors and Darshan Jain, 2023)

stage intelligent automation programs must coordinate and integrate with the broader digital transformation agenda of the organisation. And therein lies the blind spot for organisations seeking to become digital businesses.

The Blind Spot in Digital Transformation

The Blind Spot metaphor is too good to resist, and indeed much can be made of executive and organisational 'blind spots'. For example, in her 2007 book, *Blind Spots,* Madeline Van Hecke devotes a chapter to each of ten mental blind spots that afflict even the smartest people: 'not stopping to think;' 'jumping to conclusions;' 'my-side bias;' 'getting trapped by assumptions and categories;' 'not paying attention to details;' 'not understanding themselves;' 'ignoring evidence;' 'missing hidden causes;' and 'not seeing the big picture.' These may sound all too familiar!

Our research discovered a blind spot in many digital transformations. The blind spot arises—especially in very large organisations—when those driving the digital transformation agenda (the 'Strategists' in Fig. 17.1) have different agendas, resources, stakeholders and organisations than those responsible for achieving the automation of knowledge work.

Typically, 'DT Strategists' are doing a lot of right things. Especially, they develop good digital strategy and planning processes, and their digital platform is benefitting from a lot of care and attention, but their execution of strategy is weakened through a variable mix of governance, navigation, culture and change management factors. Most typically 'DT Strategists' are driving their agenda from a higher position in the organisation than the 'Automation

Technologist' stakeholders, who are focused on scaling automation infrastructure to create an imbedded enterprise capability, typically governed by an automation Centre of Excellence (COE).

All parties may feel success, but in practice the organisation has created new silos for old. The result: disconnects—both technological, and organisational. Technologically, as we saw in Chapter 16, there can be at least ten major technologies in play for true digital transformation. Intelligent automation makes inroads into converging only a few of these—typically analytics, cloud, automation, and, to some extent, Internet of Things. Organisationally, the processes and staffing for the DT and automation agendas are separate, and do not converge, resulting in a 'ships passing in the night' outcome. *'DT Strategists' view automation as a tactical tool; 'Automation Technologists' see digital transformation as little to do with them.*

Automation Execution: The Intel Inside

How to join the elite group of high performers? Well, undoubtedly automation execution capabilities are essential to digital transformation. Their DT role reminds us of 'Intel Inside' which appeared on over three quarters of the world's PCs. It meant that the computer/laptop had an Intel CPU inside. Without stretching the analogy too far, intelligent automation—the convergence of robotic process and cognitive automation technologies—also has a fundamental role, but this time in delivering something much bigger: digital transformation.

But let's start where we find most enterprises today. Typically, they will have built a Centre of Automation Excellence. They likely have between 50 and 100 digital workers, maybe even a lot more, and are getting strong **Efficiency** gains. For many, further investment looks costly, while big gains from scaling within functions and across functions look hard. Others have made more inroads into intelligent automation. Recall from earlier chapters that we define this as the growing practice of combining RPA and cognitive automation technologies and majoring on machine learning; algorithms; statistical techniques; image processing; natural language processing; and advances in computing power and memory. For clients this is regularly achieved through some sort of supplier digital exchange platform, providing tools for an ever-evolving, smarter digital workforce, often giving significant business advantage—especially when integrated with other digital technologies such as business analytics, cloud computing, and Internet of Things.

All this is essential work, and organisations must continue to improve their automation execution capability. What does this look like? Here we borrow from an expert service provider. Based on two decades of experience, Blue Prism has identified five core 'build' capabilities and 22 underpinning skills sets that maximise automation payoffs. These are organised as:

- **Pipeline**—Intake approach; Opportunity assessment; Pipeline reporting
- **Delivery**—Delivery methodology; Define; Design; Build; Test; UAT; Deploy;

  Operational support

- **Service Model**—Business as usual; Support model; Program/ROI metrics; Business reporting
- **Technology**—Security; Application management; Technical infrastructure; Platform maintenance
- **People**—Organisational structure; Capability; Training

These capabilities, skill sets and technologies provide the 'intelligent automation inside' essential for Digital Transformation. How to gain the further exponential value available?

The Secret of Success: Integration

Happily, we have seen a few organisations break through this impasse. As suggested in Chapter 16, the integration trend is likely to grow, and cannot remain an outlier. Two illustrative examples of success, can be taken from the financial services industry. In 2015, one North American bank adopted a new value-oriented, purpose-driven management philosophy of increasing organisational agility and improving customer experiences. A key focus involved transforming disjointed operating processes on an end-to-end basis, but from the customer's perspective. The automation business case was based on increasing the value of the bank's services as measured by customer metrics—retention rates, service expansion, and improved net promoter scores—rather than simply 'doing (bad) things faster'. In addition to **Efficiency** savings estimated at more than 200 percent from the ability to access and use previously trapped data, the bank also estimated a 400 percent gain in enterprise **Effectiveness**—measured by increased customer retention and revenues from broader services integration. The bank's Intelligent Automation platform has also supported greater **Enablement** gains in terms of new products and services, enterprise resilience, first-mover advantage, public goodwill and reputational equity. The bank estimated the resulting gains in enterprise **Enablement** to be greater even than the combined **Efficiency** and **Effectiveness** gains.

Similarly, a major Middle East bank undertook an enterprise-wide transformation to seize a leadership position in its key markets, using IA technology as a strategic platform. By combining Intelligent Automation with Natural Language Processing (NLP), machine learning (ML), and data mining tools, the bank developed a totally automated end-to-end solution to track payment status, pull relevant payment and customer details, and apply rule-based validations, referrals and query responses, with no manual intervention. The solution delivered 100 percent improvement in quality, response times, and customer experience. The new process also generated **Enablement** gains from the resulting wealth of data and management information—raw material for applying data science using Hadoop to improve management decision-making. International transaction enquiry handling turnaround time also went from eight to twelve hours to less than two minutes, resulting in a transformed customer experience.

The senior executives in these organisations adopted a strategic mind-set in deploying Intelligent Automation. With a transformative view and an enterprise vision suffusing from the top, the strategic uses of automation delivered greater **Effectiveness** and **Enablement** gains, beyond just **Efficiency**. These organisations also started with an external focus on customers and competition, using that perspective to design—from outward in—an end-to-end business process architecture that accelerates digital innovation. The banks' experiences demonstrate that building a robust in-house automation capability creates flexibility and a knowledge base which, with strong governance and disciplined behaviours, form part of the **Enablement** platform that accelerates strategic uses of automation technologies.

Clearly these organisations are not just moving rapidly across the horizontal axis of Fig. 17.1, but also up the vertical axis, putting themselves on the edge of being digital leaders. But they have a long way to go to match Singapore-based DBS bank, several times voted the world's best bank, and an exemplar of digital leadership in our research (see Chapter 22), building over time all the required digital transformation capabilities, and deploying intelligent automation well integrated into their digital platforms.

Towards Digital Leadership

Recently we have seen more digital leaders emerge. Here are two examples:

Using KCP diagnostics, one bank rated their Automation Execution capability and Digital Transformation capabilities at around 70 percent. In practice, all their capabilities needed strengthening, but the diagnostic provided a disaggregated analysis and pointed them precisely to where they need to focus and improve over the next year within each capability.

Automation Execution capability consists of five capabilities: pipeline, delivery, service model, technology and people. On Automation Execution maturity, the bank identified weaknesses in applications management and platform maintenance. More work needed to be done on service definition, design and delivery methodology, and also, in the people capability, specifically on training.

Digital Transformation maturity is measured as progress on seven capabilities: strategy, integrated planning, governance, imbedded culture, digital platform, change management and navigation. We provide more detail on this in Chapter 18. On Digital Transformation maturity, the bank proved very strong on strategy and planning, weaker on governance, change management and especially navigation. The analysis enabled it to set itself ambitious improvement targets for 2022 and 2023.

A second major bank has a digital workforce numbering many hundreds and had a score of over 70 percent on Automation Execution capability, but in the lower 60s percent for Digital Transformation maturity. This put them in the Digital Leadership box (Fig. 17.1)—a pleasant surprise as they felt day-to-day more like firefighting technologists! That sense arose because they were indeed scoring themselves low on technology, and had a lot to do on technology infrastructure and applications management. Other weak points identified included ROI metrics, testing, business reporting and training. On Digital Transformation maturity they were making good progress—above the industry average—on strategy, planning and digital platform, but were scoring between 50–60 percent on culture, change management, governance and navigation. This galvanised them into ambitious targets over the next year to push them higher into the digital leadership box, aligning their Automation Execution and Digital Transformation capabilities (Fig. 17.1). The diagnostic pointed them precisely to targeting particular skills sets within each capability.

Both banks were managing well the blind spot we identified earlier, and integrating Automation Execution with Digital Transformation capabilities.

But both also recognise that they are hardly done and that digital transformation is a continuous and never-ending improvement process, if they are to outflank competitors on similar journeys.

The Progress Being Made

One of the myths perpetuated on the internet, even in articles and books published in 2023, is that *'90 percent of CEOs believe the digital economy will impact their industry, but less than 15 percent are executing on a digital strategy'*. We eventually found the source of this quote. It is from a 2012 MIT/Cap Gemini study. The quote is, of course, out of date, concurrently incorrect, and its continuous use very misleading. MIT/Cap Gemini themselves have performed many more recent studies, as have McKinsey and MIT, and their work—as at 2023—is consistent with our own recent findings.

For example, Cap Gemini Research Institute identified accelerated Digital Transformation progress over even a two-year period—2020–2022. In 2020, on average 60 percent of organisations (66 percent in banking) reported they had the digital capabilities, and 62 percent (66 percent in banking) the leadership capabilities required for digital transformation.

Looking across multiple studies, organisations generally, and digital leaders especially, have increased their investments in digital transformation since 2018, increased their adoption of emerging technologies for enhancing their digital platform, and given renewed focus on talent, culture, operations, customer experience, and business innovation. The COVID-19 pandemic greatly accelerated this shift to digital. Even so, research studies are finding that over the last five years digital leaders have been widening even further the significant pre-existing gaps between them and their competitors on digital transformation, capabilities, practices and performance.

These findings are consistent with KCP findings already published. Our most recent research has been examining how organisations can make further progress whatever their competitive positioning. As detailed in Chapter 18, the key lies in building distinctive, identifiable capabilities and integrating them to create the synergies that produce significant business value. The evidence suggests that, by 2023–2024, not doing so was no longer a competitive option.

Conclusion

Significantly, research shows that companies that have embraced the digital world and execute on their digital transformation strategy register real gains in shareholder and stakeholder value. Even in 2013 digital leaders were typically above their industry average by 9 percent on revenues, 26 percent by profitability and 12 percent by market valuation. Meanwhile, against their industry average, digital beginners were seeing up to 4 percent lower revenues, 24 percent lower profitability, and 7 percent lower market valuation. A 2018 McKinsey study, looking just at automation, predicted that leaders stood to register an annual cash flow growth rate of six percent, thus doubling their cash flow between 2018 and 2030. Meanwhile automation laggards stood to experience a 20 percent decline in cash levels over the same period.

Over the 2018–2023 period, the gap widened between digital leaders and all other organisations. As one example, a 2018 CGRI analysis found a 38 percent difference on testing digital ideas quickly, expanding to a 48 percent gap during 2020. Through 2020–2022, KCP research was finding that many more organisations were moving faster on digital, in some cases driven by the need to survive in pandemic conditions. However, digital leaders were investing larger and faster than the industry average, and more strategically. Moreover, digital leaders were improving their own execution capabilities, gaining further ground rapidly on revenues, profitability, market valuation, and other leading indicators.

Digital leaders will, by instinct and practice, endeavour to stay ahead, but the widening gap between them and the rest is not inevitable. Time and again we find that in terms of achieving superior business performance the productive combination of 25 percent technology, and 75 percent management makes the real difference (see Chapter 1). In Chapter 18 we detail the seven core DT capabilities that make the management difference.

18

Core Capabilities for Digital Transformation

A new calling can beckon us away from comfortable routine and from competencies already acquired.
Neil Maxwell

Core capabilities are a fundamental way of competing, as they differentiate an organisation by being valuable, rare, not easily replicable or substituted, and arise as distinctive and required creations of that organisation.
The Authors

Introduction

In our IT and business research over a 40-year timeframe, a distinct evolutionary pattern has emerged in technology adoption, impact and value. From Office Automation in the 1980s, through IT Outsourcing (and later BPO) in the 1990s and early 2000s, to the Internet and Cloud Computing and more recently RPA and Intelligent Automation, this pattern is self-evident and self-reinforcing: ***new technology is always initially valued and deployed as a vehicle for cost reduction.***

Whether by eliminating typing pools and secretarial staffs, externalising costly 'non-core' and specialist technology-based functions to third parties, engaging with customers and suppliers directly online, shifting on-premises IT infrastructure and business processes to the cloud, or automating repetitive back-office tasks, early deployments and investments are invariably evaluated,

L. P. Willcocks et al., *Maximizing Value with Automation and Digital Transformation,* https://doi.org/10.1007/978-3-031-46569-7_18

deployed and justified on the basis of the labour and expense savings they enable.

As technologies and experiences mature, however, the cost-reduction paradigm gives way to a much larger **innovation- and *transformation-based* value paradigm**, at least amongst savvy and ambitious buyers. These companies see new technology as offering a powerful competitive advantage—a platform for innovation and business expansion at scale: new product and service development, deeper and wider market penetration, greater agility, resilience and speed to market, higher enterprise performance, and greater customer and employee satisfaction. The problem? As discussed in Chapter 8, these advantages are nowhere to be found or accounted for in traditional ROI or automation efficiency metrics. Early adoption and learning, coupled with broad, visionary leadership and requisite navigation tools, yield outsized gains for an elite group of companies who understand, commit to and execute digital transformation.

It is worth reiterating the size of the prize. Recall Chapter 8 and also our many case examples in earlier chapters. When deploying just automation, **Efficiency** gains of 30–200 percent or more are available from 'doing things right'—greater accuracy, quality, speed, asset utilisation, and cost reduction. **Effectiveness** gains from 'doing the right things' can double such gains, for example from improved analytics, greater agility, more enterprise responsiveness, higher customer and employee satisfaction, and improved regulatory compliance. But our most recent research suggests that the total available returns from digital transformation are typically four to five times straight ROI calculations, taking into account the wider transformational **Enablement** benefits from applying automation and other digital technologies. As we will see in Chapter 20, such exponential gains come fundamentally from having a flexible, 'intelligent' digital platform that supports major transformational options. We have seen, for example, first-mover advantages including innovations in product and service offerings, greater market penetration, faster acquisitions, differentiated customer and employee experiences, all leading to significant increases in revenue and profit. Our own research mirrors earlier findings by the McKinsey Global Institute, which has forecasted that 'front-runners' in adopting intelligent automation and more advanced AI could double their cashflows over the next 8–10 years, while laggards could see their cashflows decline by 20 percent.

But if the prize is huge, it is also difficult to grasp. Adding to the challenges mentioned in earlier chapters, there are value blocks and value pipeline leaks everywhere. Up to a quarter of this lost value occurs in the planning and start-up stages—e.g., low targets, poor planning processes, bureaucratic governance. Some two-thirds occurs during and after implementation partly as a result of factors we highlighted in Chapter 17 namely:

1. Lack of discipline developing and applying automation execution capabilities.
2. Not integrating intelligent automation with Digital Transformation activities. Our finding is this: *speed to value occurs when the digital transformation and automation execution capabilities are aligned and in place early*.
3. The degree to which organisations build core capabilities for digital transformation. Our research shows that this is the more fundamental factor that explains levels of success/failure. It requires executive level engagement and serious resourcing to establish an integrated digital transformation strategy and an evolutionary trajectory over 3–5 years.

Below we show the components of a resource-based approach to digital transformation and the seven core capabilities that make the performance difference.

How to Accelerate Progress

To succeed, organisations need to develop the automation and digital technologies detailed in Chapter 16 as the foundation of a digital platform, which accelerates responsiveness and time to market; provides agility plus resilience; virtualises operating and business models offers strategic options, and produces exponential business value. Amongst a very few digital leaders, this has already been happening. But for others, as we shall see in Chapter 19, there are ways of catching up—and applying a resource-based capabilities perspective to the business is a fundamental requirement.

Resources are the tangible and intangible assets an organisation uses. A capability is a set of distinctive skills and ways of organising and working that, when applied, can convert resources into effective business activities. A capability becomes core when it is essential for organisational strategic direction and operational functioning, exists internally, and helps to define, and is distinctive to the organisation. Core capabilities are a fundamental way of competing, as they differentiate an organisation by being valuable, rare, not easily replicable or substituted, and arise as distinctive and required creations of that organisation. Based on extensive and long-term research into over 1100 organisations carried out by the London School of Economics and

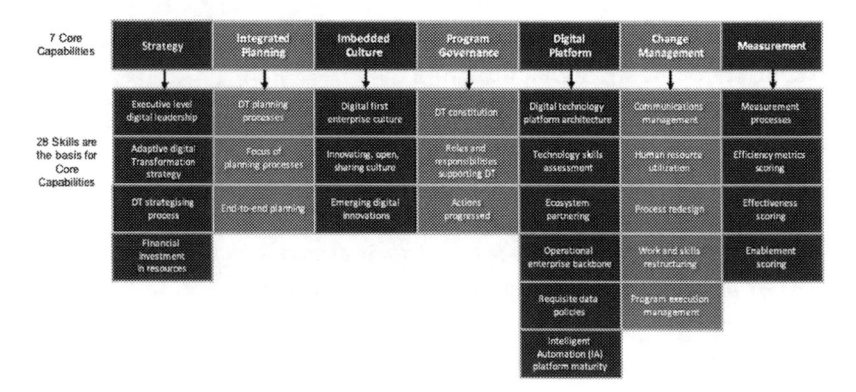

Fig. 18.1 Digital transformation capabilities (*Source* Knowledge Capital Partners, 2023)

Knowledge Capital Partners, we have identified such seven core capabilities required for digital transformation. These capabilities, suitably integrated, form a **core competence of digital transformation,** measured by total net business value gained.

These core capabilities are shown in Fig. 18.1. It is worth looking at these in more detail, as they have emerged with much more clarity more recently, and some may be unfamiliar.

Business strategy is an integrated set of commitments and actions designed to exploit core competencies and gain competitive advantage. Digital strategy is part of business strategy and establishes how to maximise the combined business benefits of data assets and technology-focused initiatives. Digital strategy sees the application of digital technologies to enable new and differentiating business models and capabilities. Central to today's organisation, digital strategy is required for its delivery of Digital Transformation. Therefore, **digital transformation maturity** reflects advanced alignment of strategy; integrated planning; program governance; imbedded culture; digital platform; change management; and navigation capabilities. Let us look at these in more detail …

1. **Strategy Capability: '*establishes the vision and direction for digital businesses.*'** This represents the ability to establish sustainable competitive, customer-focused short-, mid- and long-term business direction informed by digital technology developments, driven by business imperatives, and steered by c-suite executives pursuing an organisational transformation agenda.
2. **Integrated Planning Capability: '*details and dynamically updates how digital transformation strategy is enacted.*'** This consists of a detailed set

of organisation-wide processes designed to dynamically align strategy with operations, and coordinate developments in data, technology, processes and people to deliver a digital transformation agenda.

3. **Imbedded Culture Capability:** *'shared 'digital first' norms, beliefs and behaviours that underpin strategic and operational practices.'* The 'digital first' culture fosters constant innovation for business impact, speed in execution, openness to and sharing of diverse sources of information and insight, and allows people high levels of discretion on what needs to be done.

4. **Program Governance Capability:** *'draws up the decision-making rules and processes for digital transformation.'* This capability sees the establishment of a digital transformation constitution for how decisions get made, resources get allocated and how duties and responsibilities and deliverables are assigned,

5. **Digital Platform Capability:** *'the founding and continuous improvement of the enablement platform for digital business.'* This consists of design and development that integrates on an evolutionary basis, through architecture, the use of data and multiple digital technologies with legacy business applications in order to support business strategic direction and operational excellence.

6. **Change Management Capability:** *'the absorptive capacity of the organisation to dynamically innovate using automation and digital technologies.'* *This capability ensures that* change management integrates SMAC/BRAIDA technologies optimally with requisite developments in skills, processes, structure, culture, reward systems, management orientations, team-building and strategic direction.

7. **Navigation Capability:** *'measuring to keep strategy on course'.* This capability provides a set of processes that establishes **Efficiency**, **Effectiveness** and **Enablement** metrics (see Chapter 8) designed to monitor and apply the learning from the business value gained from automation and digital transformation activities.

The Practices of Digital Leaders

Looking across our case and survey database of 350 plus organisations undertaking digital transformation, we can say that roughly, depending on sector, some 20 percent can be classified as 'leaders'—strong in most of the capabilities mentioned above. Meanwhile some 25 percent are 'followers', 35 percent 'laggards' and another 20 percent 'nascent' organisations. Drawing on our detailed evidence, in this section we specify the practices of digital leaders. In the next chapter we will look at how other organisations can catch up. To anticipate, it is not by straight copycat replication of leader practices.

It is important to say that digital leaders are far from complacent, but exhibit continuous efforts to improve at becoming digital. Part of this has been identifying deficiencies in their DT capabilities, and working on strengthening them. Leaders also seem to be able to sustain effort and resources over much longer periods than other organisations.

What then are the characteristics of digital leaders? In practice, looking across our database as at mid-2023, digital leaders do not score high on all core capabilities but they do score high on most of them. Looking at **strategy**, sustainable competitive, customer-focused short-, mid- and long-term strategies tend to be informed by digital technology developments, driven by business imperatives, and steered and resourced by c-suite executives visioning, pursuing and communicating an organisational transformation agenda. Strategising for DT and automation is dynamic, on-going, and inclusive, with broad stakeholder involvement, communication, and education. Nearly all leaders score very high on strategy.

On **integrated planning** digital leaders, typically, have designed a detailed set of organisation-wide processes to dynamically align strategy with operations, and coordinate developments in data, technology, processes and people to deliver a digital transformation agenda. Planning focuses on building an agile, scalable digital operations platform supporting core operating systems, processes, technologies, talent and data. Planning has a strong end-to-end process and process excellence focus across and aiming to change remaining systems, functions, and organisational siloes. Once again, digital leaders typically score high on integrated planning.

Digital leaders have an **imbedded culture** consisting of shared 'digital first' norms, beliefs and behaviours that underpin strategic and operational practices and focus on continuous, technology-based renewal. There is promotion and realisation of an innovating, open, sharing culture across all management levels and functions. There is strong incubation and acceleration of emerging digital innovations for wider adoption and deployment. Here we get some variation, as even digital leaders who score high in aggregate across the seven core capabilities, are finding building the digital culture challenging and taking more time than they anticipated. DBS Bank in Chapter 22 is one high profile example of getting it right.

With **program governance**, the digital leader typically has a digital transformation constitution for how decisions get made, resources get allocated and duties and responsibilities, deliverables and rewards are assigned.

Accountability for coordinated actions and follow-up are progressed through reporting systems, program reviews and action groups. Self-confessedly, several digital leaders see room for improvement in this area, but a general finding is that organisations that get the management preliminaries right early—that is are strong in strategy, integrated planning and program governance—tend to gain significantly better returns than those who do not.

Digital platform capability represents the engine room for digital business. Not surprisingly, digital leaders that score high on this also show disproportionate returns. Several detailed examples from financial services, together with our assessments, appear in Chapter 20. The digital leader, typically, has founded and is continuously improving a technology enablement platform for digital business. There is a well-developed architecture and blueprint, and strong capability in assessment of the tools, skills and technical demands for platform design, construction and operation. The network of technology partners is requisite, updated regularly, and is utilised synergistically with internal resources. There is a strong and developing operations technology backbone in the form of corporate and service provider networks, infrastructure services, shared technology/applications, shared data and standard processes. The organisation is well advanced on integrated development, curation and ethical use of digital data. We found digital leaders, as a group more critical of their digital platform capabilities, but from an outsider perspective, this often seemed unmerited given the technological complexity, challenges and ever-changing technologies they were dealing with.

Change management receives the most variable ratings amongst digital leaders. Many find sustaining attention and resources for two years or more a real challenge. Others recognise they need to build their change management capability further. The highest scoring digital leaders integrate digital technologies optimally with requisite developments in skills, processes, structure, culture, reward systems, management orientations, team building and strategic direction. There are robust channels and programs of communication for managing change at enterprise, functional and business unit levels, including service providers. There is effective human resource organisation, including people management systems in place for skills evaluation and development utilising both internal and external marketplaces. Process excellence capabilities and tools enable redefining, re-engineering and creating end-to-end processes. Workforce management is strong, with work, skills and rewards systems requirements identified in detail. The organisation is well resourced to design and support a hybrid human–machine digital workforce. The DT program is being actioned successfully, through strong process change and effective workforce management, enabling the organisation to deal with unanticipated challenges.

We come to the final and lowest scoring capability of the seven, namely **navigation**. The lower scores across the board reflect, in our view, the general historical legacy of indifferent measurement systems mentioned in Chapter 8. Digital leaders still tend to rely on senior executives just buying into the potential for business transformation from digital technologies. They would be greatly aided by adopting better steering mechanisms to identify, track and deliver on that potential. What would this look like?

A very few digital leaders are close to the following: a set of organisation-wide navigation processes, embodying all costs and **Efficiency**, **Effectiveness** and **Enablement** metrics, are used to monitor and apply the learning from the business value gained from automation and digital transformation activities. **Efficiency** gains from the DT program are 200 percent or more. **Effectiveness** gains are 200 percent or more. As a result of the DT program **Enablement** has improved by 300 percent or more.

Conclusion

Resource-based competition, as outlined here, does present at least three major challenges. At what point do resources become capabilities, and how do we know which are core capabilities that need to be built in-house? Our research has answered this question. There are seven core capabilities required for optimal payoffs from digital transformation.

Secondly, a capability is only potential. It can only be proven in performance, and that requires integration of capabilities into a core competence, and managements capable of doing this.

Thirdly, resource-based competition involves building from the bottom up, and this may extend the time horizon in which to develop and exploit differentiation. Do today's organisations have time for this? Basically, an enterprise has to make a strategic bet on each capability. Our research, consistent with many other studies, suggests that such strategic bets are indeed worthwhile and performance differentiating, and are creating digital leaders who are pulling away from the pack.

But what happens if, so far, you have not made those strategic bets, have not called it right? The next chapter answers the still lingering question: "*If you are not a digital leader, what can you be doing?*"

19

How to Catch Up on Digital Transformation

There are no secrets to success. It is the result of preparation, hard work, and learning from failure.

Colin Powell

At least 40% of all businesses will die in the next 10 years… if they don't figure out how to change their entire company to accommodate technologies.

John Chambers

There are no traffic jams on the extra mile.

Zig Ziglar

Introduction

In the last chapter we saw digital leaders seemingly accelerating away from other organisations through developing and deploying their core capabilities for digital transformation. But there is nothing inevitable about this scenario. One mistake is to assume that the way forward is to replicate what digital leaders are doing. But this is the 'best practice' myth that has beset management consultancy advice for decades. A simple analogy exposes its limitations. If you are educating a remedial child, do you give it the same treatment as the brightest of the bright? Do you make the same assumptions about how progress will be made? No, you do not. A cruder analogy is: "*Does one shoe fit all?*" In the present case it is actually impracticable to ask what we have

L. P. Willcocks et al., *Maximizing Value with Automation and Digital Transformation*, https://doi.org/10.1007/978-3-031-46569-7_19

called follower, laggard and nascent organisations to replicate digital leader practices. Mostly they lack the resources, time, and capability to do so. But they cannot just copy what leader organisations are doing because each organisation starts from a different place, and each needs customised strategies to progress. A first step is to find out where they actually are. We provide a short assessment tool that helps this process—it is a very brief version of a much more detailed benchmarking diagnostic we use in our research and advisory work (see www.knowledgecapitalpartners.com, DT diagnostic).

Our research supports detailed suggestions as to what organisations other than leaders can execute to make significant headway. The calls to action are generalised, but designed to be still very informative to organisations identifying themselves as followers, laggards or nascent with regard to DT practices.

Follower Organisations: Making Progress

This section spells out the typical characteristics of DT 'followers'—about 25 percent of organisations, depending on sector—and what can be done to move these organisations forward. The leading indicators from Fig. 18.1 are repeated for reader convenience.

Strategy

The leading indicators:

* Executive-level digital leadership (C or C − 1)
* Adaptive digital transformation strategy driven by strategic business imperatives.
* Inclusive and dynamic DT strategising process
* Strategic financial investment in resources

Assessment: Typically, amongst followers, components of a digital transformation strategy are arrived at through a strategising process that makes limited financial investments in resources. Typical attributes are: DT initiatives defined and led at divisional/functional level; learnings and results shared on an ad-hoc basis with other business units; DT investments mostly tactical, spot-focused, near-term investments aimed at reducing cost to operate; learning from others for near-term opportunities to apply technology locally; limited central funding available on a discretionary/competitive basis with other business units.

What to do: Follower organisations are likely to be too siloed to establish centralised digital momentum and coordination. Different units will move at different paces on different agendas. They need to work more on digital governance and joined up, strategic, cross-functional projects. They tend to still have the problem of not taking a whole organisation, end-to-end process view of what they are doing with digital technologies. Greater ambition is needed on what can be achieved with these technologies, and on investing in innovation. Follower organisations feel they are making good progress but are still too functionally and cost focused and miss the exponential gains from investing in digital more broadly and strategically. Senior executives could broaden their goals and coordinate digital activities across functions. Meanwhile financial investment could be more centralised and focused on scaling and leveraging digital technologies for strategic purpose. Senior executives should address the governance issues holding this back, and through that increase focus, and redirect financial investments.

Integrated Planning

The leading indicators:

* DT planning processes
* Planning processes focused on building agile, scalable digital operations platform
* End-to-end focus across extant systems, functional, and organisational siloes

Assessment: Centrally approved DT tools and planning templates are used to plan for local operating requirements and targets. Business units can engage third-party advisors to help with local planning and implementation. There are occasional fora and frameworks for sharing divisional and functional technology strategies and plans across the business, with IT involved in evaluation and decision-making processes. Business units and functions reach out to each other when developing plans where processes intersect, but plans are primarily focused on improving functional and divisional financial performance.

What to do: Share and compare results across functional and divisional boundaries; work with HR to build and implement competency models for Centre of Excellence (CoE) staff recruitment and assessment, and with IT to evaluate third-party advisor performance against value gains. Work with IT and Operations to identify end-to-end value chain integration opportunities. Build on existing internal relationships, including IT and business planning, to identify opportunities for end-to-end process integration and improvement. Seek external advice from automation and strategy consultants on data standardisation/sharing, cross-functional investment and operational improvement. Begin to build a digital transformation culture. While it is critically important to involve IT in technology decisions, a fragmented planning approach and an opportunistic learning model limits the follower's ability to capture the full value of digital transformation. A dynamic technology environment requires sustained, regular, and comprehensive planning to achieve full performance and value. Elevate and integrate technology planning into strategic enterprise business planning.

Imbedded Culture

The leading indicators are:

* 'Digital First' enterprise culture
* Innovating, open, sharing culture across all management levels and functions
* Incubation and acceleration of emerging digital innovations

Assessment: Typically, with followers, a somewhat decentralised organisational culture encourages assessment of unit costs and opportunities to capture value from digital technologies, including where processes intersect with other business units. The culture is limitedly open, sharing and innovative. Divisional and functional technology plans receive regular management reviews to identify areas of potential cooperation and coordination. The organisation and business units look at emerging digital innovations by actively monitoring new technology developments in the wider market and work with IT to identify opportunities to improve operational performance across business processes.

What to do: Some good technical innovations are apparent, but a more organisation-wide culture supporting digital investments is needed. Senior executives need to drive a sustained cultural development effort in parallel with high profile digital 'wins'. The organisation is still at the early stages of developing a strong innovating, open, sharing culture that would promote digital technologies and enable delivery on their potential. A good tactic would be to promote and reward innovations in work processes using technology. Followers tend to recognise and support digital technologies but need to work out how governance and a broader business focus—beyond functions—will lead to bigger goals and investments.

Program Governance

The leading indicators:

- Digital transformation constitution
- Roles and responsibilities supporting digital operations transformation
- Action progressed through reporting systems and program reviews
- Automation governance maturity (ACM score)

Assessment: IT governance may be centralised or federalised but a small team in IT has a watching brief to identify digital technologies that can potentially improve operations in key business functions. A small core team works with IT in advising and sharing best practice across local centres of excellence (CoEs) in their digital and automation programs. Local CoEs have primary and autonomous responsibility for digital and automation delivery and results. There is ad hoc progress monitoring through established monthly/quarterly business reviews.

What to do: There is good prospecting, i.e., listening to the technology, but also a need to establish broader governance mechanisms with a broader range of stakeholders. For digital advances to become transformational, they have to be owned by the business with IT in support. The organisation needs to move up a level or two with more strategically focused governance mechanisms. On execution, monitoring needs to be less retrospective and more in real-time, and also tied to rewards and people responsible for results, and more immediately actionable. Are the right things being monitored—the actions that have the biggest impacts—or has this become an overly bureaucratic exercise.

Digital Platform

The leading indicators:

- Architecture and blueprint for agile enterprise digital technology platform
- Skills for platform design, construction and operation

- Ecosystem partnering
- Operational enterprise technology backbone
- Integrated development, curation and use of digital data

Assessment: The digital architecture and platform develops incrementally as business units and CoEs implement local automations. IT provides overall governance and integration with systems of record. HR and third-party advisors provide staff training and assessment, IT provides technical tool evaluation. The organisation works with various vendors and external advisors, depending on the functional processes under development. Partnering decisions are made and managed by individual business units. There are a number of local technology platforms and infrastructures. IT manages platform integration across multiple acquisitions and service providers. An overall digital data strategy and architecture is under development to provide access to enterprise data, and to ensure data integrity, accuracy and protection.

What to do: Progress can be made by introducing disciplines that tie in each project to an overall architecture and technology trajectory. There is work to do on ensuring those architectures and trajectory plans are reliable in the first place. Tool evaluation needs business users reviewing what business applications are needed and the tools that best suit. Staff training and assessment cannot be the sole preserve of the HR department, let alone external advisers. Pool skills where they are needed to deliver on more ambitious projects, and provide ongoing hands-on training. Using the external services market on a selective sourcing 'horses for courses' basis can always be strengthened by building internal capabilities that are business-, technology- and service-facing, but build a more strategically inclined sourcing function that links with and helps deliver the overall DT strategy. There are probably too many platforms, and too many applications and a blue print should be developed on how to standardise data and technologies, and limit the number of IT service providers.

On requisite data policies there are moves in the right direction, but they are the start of a 2–3-year journey, which needs focused resources to support the necessary activities. The danger is underestimating how key the right data and its organisation are to DT.

Typically, 'followers' should invest in scaling into automating other functions and processes and enterprise applications and also in building the requisite staffing and management capability for this. They should pilot combining RPA with more advanced technologies but must ensure these are synergistic linkages.

A follower tends to have a viable automation platform that need more linkages with the mainline digital transformation initiatives, or risk becoming an automation island creating a scaled automation silo which will block progress and lead to work-rounds and ad hoc solutions (see Chapter 17). Learning has to become more organisationally distributed and consolidated.

Change Management

The leading indicators:

* Communication management
* Human resource organisation and utilisation
* Process redesign capabilities and tools (end-to-end)
* Work and skills restructuring
* Program execution management

Assessment: High-level cross-enterprise communications channels are in place for broad, general communications with all levels, units and geographies. Operating divisions and functions bear primary responsibility for internal communications. Broad employee communications programs focus on building a common culture and business understanding. Local units and core functions look after recruitment, skills training and staff development, which can vary widely. Specialist external providers and consultants provide process advice and architecture, depending on the function or process being automated. The strategic importance of building and training a hybrid human/digital workforce is recognised—the requisite technological and cultural tools and systems are starting to be built. Program management skills vary widely by division and function. Their critical importance is recognised in their active development at both central and unit levels.

What to do: Embed cross-enterprise channels with strategic DT background and messaging to build wide awareness and support for the DT program—its rationale, expected benefits and strategic impact. In parallel, DT leadership should work with HR and Internal Communications to create content that 'localises' expected impacts and benefits for operating units. On human resource utilisation, it is good that there are enterprise-wide programs in place to create awareness and understanding around the DT program. Structurally, it is important to work with enterprise HR management in two key areas. First, to establish consistent CoE models, job profiles, evaluation templates and training programs for local units and functions. Followers should consider engaging specialist recruitment agencies to augment local recruitment staffs. Second, to conduct a skills assessment of the wider workforce and design education and training programs to support the transformation of work.

On process redesign, followers are already seeking best practice within areas of responsibility. The danger is that this may be limiting potential by staying 'within function'. Specialist functional experts can certainly deliver gains, but it's equally important to consider the extended value chain—not just within the enterprise, but including service providers and customers. Taking a strategic end-to-end enterprise view can unlock new sources of value.

On work and skills restructuring, followers tend to recognise they are managing two tracks—one is the software, the other is the 'human-ware', which can be a bigger challenge. But it is important to work with HR to seek out and promote digital champions—giving them leadership roles and support for training their colleagues, augmented by digital training consultants to create the training infrastructure.

On program execution, successful digital transformation requires strong and sustained leadership and governance, along with effective technical and project management support. Leading exemplars establish a C-level strategic program management office for the duration of the initiative, comprised of key decision makers, strategy consultants, HR and technical experts, along with representatives from local and functional units.

Navigation

The leading indicators:

* Metrics and continuous measurement processes in place
* **Efficiency** metrics scoring
* **Effectiveness** scoring
* **Enablement** scoring

Assessment: Typically, the follower organisation uses a combination of Return on Investment (ROI) and Total Cost of Ownership (TCO) to establish and build DT business cases. There is liaison across business units in terms of measuring joint projects across their lifecycle. As a result of the automation and DT programs **Efficiency** tends to have improved between 30 percent and 200 percent, **Effectiveness** by between 30 percent and 100 percent and **Enablement** by 5 percent to 100 percent.

What to do: ROI an TCO will help followers identify potential **Efficiency** gains, and perhaps gains in enterprise **Effectiveness**, but they will be missing out on other potential sources of value not easily quantified by these metrics. Followers too easily pass up on the strategic gains from DT. It is worth followers studying what other successful DTs have enabled, and also carry out a detailed analysis of where ROI and TCO metrics might be misleading managers into targeting only limited payoffs.

Followers, unlike leaders, often miss the fact that **Efficiency** gains from DT typically exceed initial expectations as efficiencies build on each other, often doubling the anticipated results. They grow and expand over time, moreover, so followers are advised to improve the accuracy and regularity of their measurement processes.

Meanwhile **Effectiveness** gains from DT build cumulatively and interactively, resulting from the ability to seize new business opportunities as and when they arise by having a flexible and responsive digital platform. These gains are typically not factored in ROI and TCO calculations, but can easily triple the gains from **Efficiency** alone. Followers are advised to be sure to include them in any business case.

Enablement gains from DT are typically many multiples of combined **Efficiency** and **Effectiveness** gains—as much as 10x in many cases. They arise from having a powerful, low-code digital business platform, connecting and integrating a growing human and digital workforce. Discounting the source of such major hidden, future value, as many followers do, is a serious mistake.

The heart of the problem with follower organisations is not taking value measurement seriously enough for both strategic and operational steering purposes. Flying with limited sight as it were, followers put in average rather than superior performances, and leave a great deal of value and competitiveness on the table. The recipe for moving forward is thorough measurement processes and moving to more appropriate evaluation techniques.

Laggard and Nascent Organisations: Taking the Next Steps

These organisations tend not to score well across the board on the seven DT capabilities. Rather than produce the long list of their deficiencies we will move on more constructively to establish the actions they can take to make strong progress.

Strategy

Action Planning: Typically, a laggard or nascent organisation has not invested in a digital transformation strategy , let alone the means to deliver it. Islands of digitisation provide very limited and local payoffs. Nevertheless, they should. not look to become best-in-class quickly but, instead evolve a comprehensive staged approach to adopting and combining digital technologies that—over several years—will add up to a pathway to digital transformation. The typical organisation allows limited goals and local interests to displace the more strategic centralised actions that reap the potential benefits from digital technologies. They also should link automation efforts more clearly with other digital technologies and broader business imperatives. We have seen lack of financial investment combine with lack of strategic senior executive support to strangle anything but very limited use of digital technologies. We advise these organisations to try to break out of seeing each investment as a one-off. Also, to get more ambitious on outcomes; for example, how can digital technologies address where the organisation is hurting operationally? Given skills shortages, how can work be made more meaningful?

Integrated Planning

Action Planning: Without a shared plan and roadmap, gains from intelligent automation will invariably be limited, discrete, and fragmented. This can be countered by seeking out other local and functional automation 'islands' to build relationships, share learnings and experience, and compare tools and methods. Also, by working with IT and/or Operations leadership to showcase capabilities and results to higher management levels.

Transformation depends on the risk tolerance and the understanding of functional and process leaders regarding digital capabilities and benefits, which will likely vary greatly. Gains, therefore, will remain localised, and performance will be uneven across the enterprise. There is value in working with an enterprise business planning team to highlight the value of a coordinated transformation approach. In the same way the current approach to value—capturing 'lowest common denominator benefits' and 'low lying fruit' in isolated islands of digital activity—carries great strategic risk. Look end-to-end across the value chain for more promising automation and digitalisation candidates.

Imbedded Culture

Action Planning: Typically, the organisation has many parts (and many cultures) but none—except probably the IT function—owns digital. The good news is that developing a culture around 'digital first' could solve many problems, as well as easing the path to digital transformation. However, these organisations are seriously lagging in gaining—let alone leveraging—a digital organisational culture. They therefore have to begin moves in this direction, perhaps on the way discovering that employees may well prefer an open, innovating sharing culture, and one that is informed by utilising digital technologies better. On digital innovation culture, these organisations, certainly the larger ones, also often have multiple units too localised in attention and investment to make anything other than very limited digital advances. A way forward is to sponsor more cross-boundary uses of technology, based on larger business goals.

Program Governance

Action Planning: Simply put, no central governance, no digital transformation. DT has to be driven by the business supported by IT, and governance provides the enabling structure. These organisations need to identify the program first, then build governance perhaps by piloting in one business unit, then scale governance to be increasingly centralised.

Centres of Excellence may well exist as a form of governance but offer both strengths and weaknesses. These organisations may well be taking action and building learning, but are all too often also creating new silos, and so seriously limiting what they can achieve. Stronger links with the rest of the organisation are vital for DT, so start coordinating across business and functional units and bring in more senior cross-unit executives. These organisations exhibit a lot of activity at the local level but should not mistake 'busyness' on automation and/or digital for digital transformation. The new focus should be on linking actions across units and aligning them with strategic direction, and emphasising the mutual benefits for a wide range of stakeholders across and up the organisation.

Digital Platform

Laggard and nascent organisation really do have a problem if they are without a digital technology architecture and blueprint. They need to revisit and strengthen their in-house technical architecture capabilities and begin to develop a 3–5-year plan of the technology trajectory using IT people to establish feasible technologies, and business unit staff to establish where the demand for applications is and will be.

Meanwhile lack of foresight on skills will hold back future adoption and leveraging automation technologies and their combination with other digital technologies. Pooling skilled resource more centrally with strong funding will allow skills to be focused on starting up and delivery of more potentially useful projects. More coordination between IT and business staffing and activities is typically a basic area for improvement, but also building a capability to leverage external resources where needed, and from whom internal staff can learn.

Typically, the enterprise strategy is to 'sweat' existing operational platforms (IT, automation, ERP). This may be a short-term response to larger crises and poor financial results, but sweating assets is not going to lead to leveraging the more powerful technologies, or improving and leveraging the data available. Eventually the organisation is going to have invest in these areas to make savings and improve performance much more dramatically. These organisations also tend to have indifferent data policies. Poor data collection, organisation and use are the dirty secrets of digital transformation and a key reason why some do not get very far on DT. These organisations need to start looking at what data can be standardised and used for larger organisational purposes.

Change Management

Action Planning: These organisations, on the whole, do not do internal communication well. For DT to succeed, it is essential that all employees—at all levels—understand the rationale, strategy , objectives, and expected benefits of the program. At a minimum, digital program management should provide local and functional internal communications and HR teams with background and key messages for inclusion and promotion across existing channels. Their human resource utilisation also needs improvement. Without a more strategic HR function, they find it challenging to acquire the specialist skills needed for DT, or to prepare the workforce for changes in their roles. One possible way forward is working with specialist third-party recruitment agencies to build a digital implementation talent capability. Also, HR partnering with internal communications to help employees prepare for the changes they will experience in a blended human and digital workforce.

Surprisingly, these organisations sometimes confuse automation and digitisation with transformation. But, to coin a phrase, 'no process is an island'—it is constantly receiving inputs and generating outputs. DT is an opportunity to re-think how existing processes work end-to-end. Therefore, it is useful to consider each process in the context of the wider enterprise value chain: what inputs and actions do they actually require, and what added value do they actually generate? Otherwise, the risk is constantly running to catch up.

On work and skills restructuring, in the mid-2020s war for digital talent, these organisations risk permanently falling behind the competition, in being late or even unable to create a competitive digital operation. In many ways this may be a more serious issue than technology per se. They should consider seeking external training or service partners for key roles, and alert HR and senior executives as to the urgent actions needed.

On program execution management, these organisations typically have an indifferent overarching enterprise change program and communications on round DT. The obvious fragmentation will lead to failure if individual business units are left to their own devices, decisions and competing priorities, and if the workforce is not adequately informed and motivated. Laggard and nascent organisations have to move to a more structured enterprise approach, as suggested for example in Chapter 21.

Navigation

Action Planning: ROI tends to be the primary financial metric to assess automation and digital technology investments, and establish business cases on a project-by-project basis. But efficiency (doing things better) is the smallest value 'bucket' available from these technologies and the measurement systems employed miss most of the value potential. It's easy to be seduced by metrics. Limited metrics lead to limited ambition, which leads to limited gains. In the world of digital transformation, '***not everything that matters get measured, and not everything we can measure matters.***' These organisations need to look for examples and outcomes that demonstrate wider business gains, beyond simply cost reduction. And consider the cost of NOT succeeding with DT. Adopting more thorough measurement processes and more suitable technology valuation frameworks would be important steps in the right direction.

Conclusion

Our digital transformation research exemplifies the words of an academic colleague at MIT George Westerman:

> When digital transformation is done right, it's like a caterpillar turning into a butterfly, but when it's done wrong, all you have is a really fast caterpillar.

Becoming a digital business requires transforming the business. This is accomplished by deploying the seven core capabilities, integrated with intelligent automation inside. Detailed research has established the robustness of the core capabilities for digital transformation shown, previously, in Fig. 18.1, and how strongly they correlate with business value payoffs. Therefore, it is worth digging deeper into some of them.

Earlier chapters have already looked in various ways at strategy, planning, governance and navigation. Ensuing chapters deal with the digital platform (Chapter 20) and culture and change management (Chapter 21) while Chapter 22 brings all the themes together with the DBS bank example.

20

Digital Platform as Foundation

Developing IT infrastructure at Macquarie bank was described to us in 2002 as trying to change the tablecloth without disturbing the cutlery. Today, in the 2020s, building a digital business platform means changing most of the cutlery as well.

Leslie Willcocks and John Hindle

Introduction

Digital transformation needs, firstly, foundations. To succeed in the emerging digital economy, organisations will need to deploy intelligent automation as the foundation of a digital enterprise platform, to accelerate responsiveness and time to market, galvanise customer experience provide agility and resilience, virtualise their operating and business models, offer ongoing strategic options, and support business value creation, at speed and scale. Amongst a very few digital leaders, this has already been happening. But for others, as we shall see, there are ways of catching up, by accelerating the process.

In Chapter 18 we identified seven essential capabilities for digital transformation. While all seven are critical for success, Strategy and Digital Platform are critical requirements at the outset of the transformation journey. Today aligning Strategy and Platform is not enough; they must be developed dialectically. Let's look at this proposition.

© The Author(s), under exclusive license to Springer Nature Switzerland AG 2024
L. P. Willcocks et al., *Maximizing Value with Automation and Digital Transformation*,
https://doi.org/10.1007/978-3-031-46569-7_20

The Strategy-Platform Dialectic

Strategy Capability establishes and dynamically updates the vision and direction for the digital business—its competitive differentiation. It defines products and services, markets, customers, channels, as well as the resources (including talent, financial, technological, material) needed to enable the strategy. It incorporates the digital strategy—the key technologies that will enable new and differentiating business models and capabilities.

Extending other studies, a **digital platform** is a repository of business, technology, and data components facilitating rapid innovation and enhancement of digital offerings for customers experiences and operational efficiency. The raw material of digital offerings and processes is a set of software components. To facilitate development of both new and enhanced offerings and processes, organisations need robust platforms that make reusable business and technology components available for reconfiguration. Amongst digital leaders, modularity is a constant design imperative.

Many practitioners and vendors define such platforms in ways that give them limited scope e.g., a sales platform, a procurement platform. In our macro view a matured digital platform can deliver transformed customer experiences, performance jumps in operational efficiency, and be the foundations for multiple interlinked business platforms. In banking, DBS bank is an exemplar of delivering such an ambition (see Chapter 22).

Digital Platform Capability sees the design, building and continuous improvement of the technology platform for strategy realisation. The foundation of the Digital Platform is a robust intelligent automation infrastructure for executing operating processes. It consists of enterprise-grade RPA platform, augmented by an ever-expanding range of AI tools and capabilities—machine learning (ML), natural language processing (NLP), optical character recognition (OCR), decision agents, etc.—connecting and unifying a diverse range of internal and external human and digital resources (employees, customers, suppliers, partners, data resources, etc.) to enable and deliver the strategy. But the digital platform is developed by further combining and optimising emerging digital technologies for business purpose.

The goal is a business-led modular technology and data platform. Our research is consistent with other studies: putting in place a fit-for-purpose modern technology architecture driven by business needs and strategic imperatives is one of the top success factors, enabling secure, scalable performance, rapid change and deployment, together with reliable ecosystem integration.

In dynamic technology and business environments, moreover, the ability to continuously reconfigure and align enterprise capabilities and resources with commercial opportunities is key. But there is no endpoint. And strategy-platform alignment must now become fusion. This requires a continuous and conscious 'dialectic' between Strategy and Platform—Strategy informs and shapes Platform choices, and Platform enables and shapes Strategy options. Platform, in effect, becomes the physical and conceptual instantiation of Strategy. And in a digital operating environment, both Strategy and Platform must be dynamic and open-ended, not fixed and immutable.

We have observed much of this amongst digital leaders. Furthermore, when making platform choices and decisions these leaders consider two frames of reference and participation:

1. **Internal**—an agile platform infrastructure that enables and accelerates continuous development and innovation by standardising enterprise component interfaces and interactions across operating functions and management entities—e.g., common data units/formats, hardware interfaces and software APIs for building applications in a 'building block' model, including enabling rules for access and utilisation.
2. **External**—a multi-function platform infrastructure that enables secure direct and third-party integrations with multiple entities and resources in the 'extended enterprise' (customers, partners, suppliers, developers, influencers, etc.) across diverse media, data types and content sources.

The key characteristics and performance dimensions that leaders consider and evaluate when building a foundation platform for digital transformation include:

- **Compatibility**—backwards/forwards, internal/external, 'open' across multiple sub- and third-party platforms—the ability to accommodate relevant protocols and interface standards for various data entities, content and media types (industry-specific or media-specific).
- **Extensibility**—the ability to scale (up or down) easily and cost-effectively, at both hardware and software levels, in near-real time, to create and respond to demand; cloud is an obvious strategy/solution.
- **Control**—the ability to monitor and manage access and performance actively across all elements of the extended platform—technology, data, security, even human resources.
- **Congeniality**—ease of use, both at the 'builder' level (designing and building solutions) and the 'customer' or 'consumption' level (users, partners, suppliers, etc.)

In addition to these technology management characteristics, there exists a third digital platform level of talent development—acquiring, developing, and training the human software needed to build and run the above. Human talent is integral to the digital platform.

Let us assess how some organisations have applied these principles in the context of Digital Transformation, with varying degrees of success, and how Strategy and Digital Platform have shaped the other five transformation levers.

Making Platform Progress in Banking and Financial Services

Without strategy, planning, governance and navigation execution is blind. Without digital platform building, and culture and change management there is no delivery, and ultimately no digital business.

Strategy listens to the technology and establishes what is possible. Planning provides the route map, governance the decision making and navigation the steering. Culture change provides the supportive norms values and behaviours, change management the execution, while the ever-maturing digital platform becomes the engine room of the digital business. Here we focus on digital platform evolution in action.

The earlier financial services companies got into computing, the more complicated their digital integration challenges today. Banks have lived through and embody multiple generations of technologies—from mainframes, desktops, client server, and internet-based technologies, and applications, not least varieties of enterprise planning database and cloud applications. Becoming digital, each bank has its unique starting points, and each its distinctive journey. We see lots of impressive digital strategies. Then we look at what passes for the digital platform and know it's not going to go too well. This is not surprising. The integration challenges are formidable. But there are ways forward. Each has to build the same seven core capabilities for digital transformation to happen. Let's see what can be done by comparing two banks at different stages, and their ways forward. Our focus is digital platform capability (for the relevant skills sets see chapter 18).

BANK 1: Established—But a Long Way to Go

Here strategy is running ahead of planning and governance. Culture and measurement are weak, but change management is doing a relatively good job with what it is being given to do. The usual barriers in this sector apply, making organisational transformation hard going: silos in strategy, structure, culture, skilling, processes, and managerial mind-sets. But what about the digital platform?

The digital technology platform architecture and blueprint develops incrementally as business units and COEs implement local automations. IT provides overall governance and integration with systems of record. The organisation needs to bring in disciplines that tie in each project to an overall architecture and technology trajectory. There is work to do on ensuring those architectures and trajectory plans are reliable in the first place.

Technology and skills assessment. HR and third-party advisors provide staff training and assessment, IT provides technical tool evaluation. Tool evaluation needs business users reviewing what business applications are needed and the tools that best suit. Staff training and assessment cannot be the sole preserve of the HR department, let alone external advisers. Pool skills where they are needed to deliver on more ambitious projects, and provide ongoing hands-on training.

Eco-system partnering. The organisation works with various vendors and external advisors, depending on the functional processes under development.

Partnering decisions are made and managed by individual business units. The bank needs to strengthen its internal partner-facing business-, technology- and service-focused capabilities. It needs a more strategically inclined sourcing function that links with and helps deliver the overall DT strategy.

Operational enterprise technology backbone. There are number of local technology platforms and infrastructures. IT manages platform integration across multiple acquisitions and service providers. The bank has too many platforms, and too many applications and needs to develop a blue print on how to standardise data and technologies, and limit the number of IT service providers.

Requisite data policies. An overall digital data strategy and architecture is under development to provide access to enterprise data, and to ensure data integrity, accuracy and protection. These are moves in the right direction, but they are the start of a two-to-three-year journey, which needs focused resources to support the necessary activities. It is dangerous to underestimate how key the right data and its organisation are to digital transformation.

Intelligent automation execution. The bank has advanced automation execution capability, but is only slightly above the sector average on overall digital strategy maturity. Clearly there is a technology disconnect here, and an example of the 'Blind Spot' highlighted in Chapter 17. The key is integrating governance, planning and financing, leading to technology integration.

Interestingly, there was quite a lot of disagreement amongst senior executives as to the state of progress in these six areas, but especially on data policies, indicative of weaknesses in planning, governance and measurement. Nevertheless, ambition was high in aiming to get the digital platform, and most of the other core capabilities, from an average to an advanced state for the sector within a year.

BANK 2: Advanced with Even More Ambitious Targets

In 2023 this bank was amongst the sector leaders on digital strategy, was advanced on integrated planning, change management, embedded digital culture and digital platform, and average for program governance and measurement.

Digital technology platform architecture and blueprint. The bank has moved from developing an architecture incrementally with overall governance and integration by IT towards a well-developed architecture and blueprint for an agile enterprise digital technology platform. In this they were ahead by probably 60 percent of the organisations in their sector. The potential problem is becoming wedded to the existing plan and technologies while new technologies and uses are emerging rapidly and continuously. They needed to ensure that design of the technology platform is open and can incorporate new technologies quickly.

Technology and skills assessment. From relying on HR and third-party advisors for staff training and assessment, and on IT for technical tool evaluation the bank is moving towards having strong central capability in assessment of tools, skills and technical demands for platform design, construction and operation. The bank will be able to drive much progress with these in place; the problem will be retaining skilled personnel, paying them enough, and providing sufficient interest. Ambitious projects attract high performers, who need to be serviced and supported well.

Ecosystem partnering. The bank is less good at working with external technology suppliers. It works with various vendors and external advisors, depending on the functional processes under development. Partnering decisions are made and managed by individual business units. Using the external services market on this selective sourcing 'horses for courses' basis needs a more strategically inclined sourcing function that links with and helps deliver the overall DT strategy. Digital transformation needs a strong internal technology sourcing capability to build a network of technology partners that is requisite, updated regularly, and is utilised synergistically with internal resources.

Operational enterprise technology backbone. This is also a weaker component. The bank has been operating across a number of local technology platforms and infrastructures, and has been relying on IT to manage platform integration across multiple acquisitions and service providers. As with many others in financial services, this bank still has too many platforms, and too many applications and really do need to develop a blue print on how to standardise data and technologies, and limit the number of IT service providers.

Requisite data policies. Unlike many we have researched this bank is relatively strong here. Having developed an overall digital data strategy and architecture to protect and provide access to enterprise data, and to ensure data integrity and accuracy. The bank is well advanced on integrated development, curation and ethical use of digital data. In this respect it has made good progress in an area most executives overlook and neglect. But the bank needs to keep focusing on data policies as a core activity rather than shifting focus on something that seems to need more urgent attention.

Intelligent automation execution. The bank is advanced, especially on delivery and service model, but is not yet a leader in its use of intelligent automation. The main catch-up work needs to be done in the areas of pipeline, technology and the pace of development here and the ambitions over the next year are aligned with overall digital strategy maturity.

These two banks are amongst the top 30 percent globally in their digital transformation efforts. They demonstrate serious work-in-progress driven by ambitious targets. Technology transformations are notoriously difficult and complex. In the large, both have avoided the top mistakes of being too piecemeal and limited in scope, not connecting technology to business value, and finding transformation too expensive to sustain. Both are becoming much more future ready in their management approach. What does this mean?

Future-Ready Digital Platforms

Facing today's environmental dynamism, connectedness and uncertainty, together with rising, increasingly unpredictable competition, organisations need to become not just digital but also agile and adaptable—that is, 'future-ready'. As several commentators point out, this means building an ambidextrous organisation and digital platform for both **exploitation**—efficient leveraging of existing resources and capabilities through continually improved processes—and **exploration**—combining resources and capabilities in new ways to create further capabilities and opportunities. Tensions between exploitation and exploration will occur, and will need to be managed by organisational integration mechanisms, including, for example, integrating a stand-alone digital bank back into the larger traditional bank.

The crucial point is that organisational ambidexterity is not enough. To be future-ready, business ambidexterity must have digital foundations, and operate through data and digital processes. MIT academics Jeanne Ross and colleagues have put it another way: in designing for digital there is a simultaneous need for a digital componentised operational backbone (to underpin exploitation), and for digital services platforms (for speed, flexibility and experimentation i.e., exploration). And, of course, this design logic of digital ambidexterity is increasingly facilitated by today's multi-layered technology architecture, supported by modularised components, API-enabled connectivity, and scalable cloud usage.

Tensions within the digital platform between exploitation and exploration can be managed through a multi-layered technology architecture with an API-based integration layer that connects core infrastructure with customer facing services and content.

The overall vision here is of a business-IT-digital fusion, with business IT and digital strategies highly interlinked, and with technology increasingly driving business innovation. In a step towards being even more future ready, some suggest that tensions in bridging Business and IT can be managed by reorganising around **business capability platforms** run by autonomous cross-functional teams and coordinated by automated processes and platform standards. However, most organisations will not be in a position to make this step yet. So, what can they do be doing?

Getting To 'Future-Ready'

Looking across our research and composite sources, organisations that are building 'future-ready' can be found working on three components. The first is a reimagined role for technology that is tightly focused on and fused with the business strategy. The business-technology gap has been a perennial issue in transformation initiatives, and has been closing slowly, over several decades. But now, in a very real sense, *digital is becoming the business*. Therefore, what has been called 'tech-forward' business strategy is critical, together with integrated bus-tech management and stewardship of the digital user experience.

Secondly, banks need to continue to reinvent technology delivery. This means agile-at-scale software delivery, modularisation, adoption of next generation infrastructure services (e.g., cloud, end-to-end automation, NoOps, PaaS), technology skills excellence (internal and external), and flexible technology partnerships.

The third component, supported by our recent research, is future-proofing the foundations—that is, the digital platform. The target state here is, firstly, a flexible, business-backed architecture developed iteratively and continuously to renew core systems so they support new digital functionalities. The process will see multiple, even daily, production releases, and frequent upgrades. The flexible architecture will consist of self-contained applications connected with easy-to-configure application programming interfaces (APIs). Secondly, the technology core will include data and analytics systems that provide technology teams across the enterprise with the high-quality information and tools they need to anticipate customer
and employee preferences, design innovative applications, and enrich user experiences. Thirdly, security and privacy protections need to be integrated into solutions from the start, rather than added subsequently. This approach can accelerate delivery while improving information security.

For digital platforms, there are many routes to future-ready. A 2021 MIT study by Stephanie Woerner and Peter Weill found only 22 percent of companies future ready, and identified four pathways to improvement that were particularly effective, producing an average net operating margin 19.3 percent points higher than the industry average.

1. Where customer experience was uncompetitive, organisations first focused technologically on creating an integrated customer experience and better digital offerings. MBank in Poland did this, then needed operations to

get up to speed, so implemented a new customer-at-the-centre banking platform.

2. Organisations with poor operational efficiency focused first on industrialising their technology platform, then on further improving the customer experience. Danske Bank and Commonwealth Bank of Australia adopted this approach.

3. A third pathway took small 'balancing steps'—alternating focus on operations and customer experience—over several iterations. A key success factor is to have an overall transformation roadmap that informs each of the separate efforts. Without this coordination, digital initiatives tend to lose their way, and the overall digital transformation project loses momentum.

4. A fourth, more radical pathway is to start a new future-ready company with a digital platform and all the advantages built in. This is a sensible option if the fintech competition is fierce, and management cannot see how to change the culture, customer experience and operations fast enough to survive.

Interestingly, Bancolombia chose multiple pathways, internally transforming customer engagement and operations systems but also creating its own digital-only banking entity—Nequi—as a test bed and internal competitor. By late 2021 80 percent of Bancolombia's own customers were digital users, and Nequi was made a separate business to grow and exploit its technology.

Conclusion

In our analysis we have pointed to the fundamental importance of getting the digital platform foundations right, and identified how progress can be made. All the research points to how difficult the overall task is, with serious disruptions and question marks along the way. Nevertheless, just looking at the banking sector as an example, some 55 percent of banks were making serious inroads into building their digital foundations by 2022. That, of course left some 45 percent of the banking sector not having made serious progress out of their technology silos and spaghetti legacy accumulated over several decades. In today's financial services, as in most other sectors, value migration is constant. And when the game changes, the winners are, like Wayne Gretzky pursuing the ice hockey puck, the ones who migrate to where the value will be.

21

The Heart of the Matter—Effective Change Management

At the heart of change is a change of heart
Leslie Willcocks

Introduction

Major projects mean big change, to which the organisation itself needs to adapt. Moving into American or Asian markets, embedding digital transformation, rolling out a new world car—such strategic moves need not just change management underpinning the project, but also require the organisation itself to change, to some degree or other. How to achieve this? Adopting program management offices and applying big project management disciplines help here, at the project/multi-project level. But to behave strategically, over three years to five years, to change culture, work habits, redesign processes, and apply and evolve digital technologies as they become available—all this does require more transformational practices.

The more strategic and transformational the project, the more a comprehensive, systematic approach is needed. As we saw in Chapter 18 culture and change management become core management concerns. Our own approach at Knowledge Capital Partners is shown in Fig. 21.1. While there are always barriers to change, few organisations take a sufficiently holistic approach to dealing with these barriers and mobilising the organisation. Ralph Kilmann did great work on this and pointed out that that there are two types of barrier

L. P. Willcocks et al., *Maximizing Value with Automation and Digital Transformation*, https://doi.org/10.1007/978-3-031-46569-7_21

that managers can do little about. One is the setting—the external environment, its complexity and external stakeholders. The other is the psyche of people—their innermost qualities that translate into their fears, desires, what they will resist and defend. The psyche cannot be changed in a short period, if at all. Drawing on Kilmann's work, the good news is that all the other likely barriers, including culture and assumptions, are manageable and changeable, provided an integrated, planned approach to change is adopted. Kilmann also makes the essential point we have discovered in all the major IT-based projects we have encountered over the last thirty years: for optimal outcomes, cultural change has to be addressed from the beginning, and throughout, not at the end of change programmes.

Culture can be thought of as the long standing, tacit but very 'stickable' habits permeating how people think, behave and deal with problems. As such culture is quite difficult to address. Historically, many have pointed to the abiding power of culture in organisations, nicely summarised in a formulation attributed to Peter Drucker as: "*Culture eats strategy for breakfast*", or quoted by Thomas Lloyd as "*culture beats strategy*". Our own studies point to the fact that, without accompanying cultural change, major change— a new product, a new, shared service centre, a new technology, digital transformation—invariably disappoints.

Figure 21.1 offers a way through the inherited, very often siloed organisation. As we mentioned in earlier chapters, silos build up in organisations in eight major areas—processes, data, technology, strategy, structure, skills, managerial mind-sets, and culture. That is a lot to change! How to do so?

There are six tracks that need to be managed on a mutually informing basis. These are the culture, management skills, team building, strategy-structure, technology-process, and rewards tracks. Managing these six

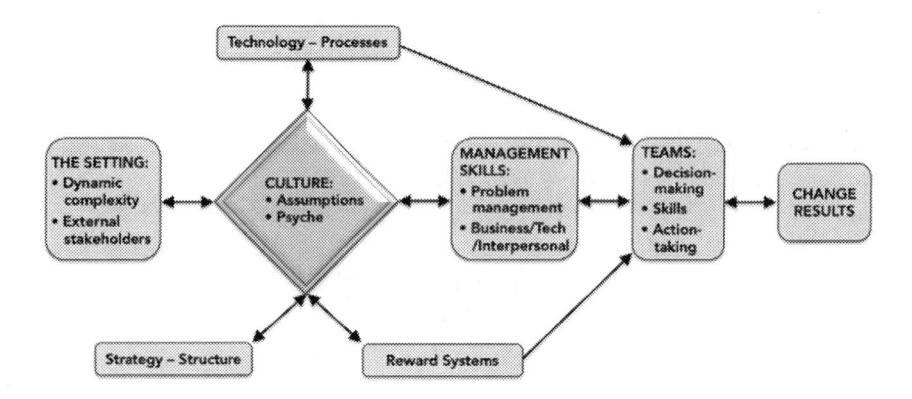

Fig. 21.1 Digital transformation starts with cultural change

tracks—starting, sequencing and keeping them mutually informed and aiming at the same target—is the province of the core capabilities of imbedded culture and change management.

How to sequence the tracks? While listening to what the technology can do, by all means build carefully crafted details of the target **strategy, structure, technology, and processes**. But these must be provisional, because the first three tracks that need attention are culture, management skills and team-building. These adjust the *behavioural* infrastructure of the organisation. Note that the focus here is on developing **an adaptive inner organisation**. Without this, changes to the outer organisation—strategy, structure, rewards systems and technologies—will be cosmetic and short-lived. However, there must be from the start a very strong sense of how all these factors fit together. For example, it is not enough to have a business strategy without also at the same time establishing how digital technologies can leverage, or even change the strategy, and whether a supportive culture is feasible. If the business strategy is based on false assumptions, then all digital technologies can do is to provide an efficient underpinning for what might be tantamount to 'disaster faster'. As in the case of DBS bank (see Chapter 22), with large-scale transformation programs, implementing and institutionalising the changes may well take several years.

The rough sequencing of the change activities is shown in Fig. 21.1. Once the assumptions on which the business decisions are based are checked to be realistic, the starting point has to be whether the culture will support the behaviour needed for organisational success. If not, shifting the culture—by identifying the shared norms, values, and change propensity organisational members are willing to commit to—must start immediately and continue throughout the change process. What should this culture be? One that stakeholders will buy into. Our own work on digital transformation suggests that having a **digital ready culture** accelerates the adoption of automation and digital technologies. George Westerman and his colleagues looked at over 500 companies in digital transition. They concluded that the four founding values of digital culture are:

1. **Impact**—changing the world radically through constant innovation;
2. **Speed**—moving fast and iterating rather than waiting to have the answers before acting;
3. **Openness**—engaging broadly with diverse sources of information and insight. Sharing advice and information openly rather than keeping knowledge to oneself;

4. **Autonomy**—Allowing people high levels of discretion to do what needs to be done rather than relying on formally structured coordination and policies.

KCP work confirms these values but identifies a further fifth vital value— a **'digital first'** culture that (1) focuses on generating technology options for business problems; (2) listens to the technology; and (3) invests in digital skills development.

But it is practices that bring cultural values to life. Therefore, in Fig. 21.1 we include those digital development and operations (Dev-Ops) practices where values are made real through, for example, rapid experimentation, time-box philosophy, data-driven decision-making, multi-functional teams and self-organisation. The difficult balancing act is to develop this culture while preserving essential, though more traditional practices, for example a primary focus on customers and results; acting with integrity; seeking stability, while shifting rules-based cultures to results-based.

The next phase is building the necessary **management skills, then teams** to deliver the change. At this point the originally formulated business strategy and organisational structure will need revisiting. On **technology**, our experience with IT and digital technologies is that these need to be in the 'first order thinking' done at the time of, and included, in the first formulation of business strategy. A feasibility analysis of organisational readiness to accept and deliver the technological changes should be carried out then, and a natural checkpoint for digital transformation comes when trying to establish stakeholder buy-in and governance. Management and teams will need to regularly liaise over technology delivery and driving out business value. Finally, the **rewards system** will need adjustment to incent changed behaviour and target required business outcomes.

Those managing these six tracks put into action the core DT capabilities of imbedded culture and change management but there has to be a seventh 'shadow' track because politics breed in times of technological change.

The 'Shadow' Politics Track

Today most technology projects of any size are seen as business not technology projects. That has become a commonplace statement to the point of cliché, but what does it mean? Such projects must be managed by a senior business manager, even the CEO who provides the vision (the 'what'), the resources, and protects the project. A senior credible, influential project champion is required who will spend 60 percent or more of their time keeping the project on track from a business perspective. Then there will be a credible, experienced project manager—perhaps several—and project teams of dedicated full-time users, technology specialists, contractors and part-time user managers and subject matter experts. Such project groups are weighted heavily towards business imperatives, and providing business knowledge, but also, inevitably breeds their own politics (that need to be managed). Even before a project group starts its work, it become clear we need change agents on board who are sensitised to the politics of the intended changes. What is true of a major technology project is even more valid for a whole organisation change like digital transformation, where the vocabulary has to be modified from 'project' to program (see Chapter 18).

This bears further consideration. The wider stakeholders in the organisation, together with external stakeholders—for example suppliers, customers, regulatory bodies—will have their different perceptions, understandings, interests, and expectations. Organisational politics DO breed in times of technological change. The extent of political activity during DT execution will depend on how far power has been mobilised behind the program, and whether other options and paths of resistance have been closed off. Political activity will rise with unclear goals and outcomes, different goals, the need to allocate limited resources, differing definitions of organisational problems and differences in information made available.

Just ponder the implications of automation and digital technologies for an organisation. New technology will release political energy because of its anticipated and actual organisational and social impacts and the manner in which it is designed and implemented. A political approach may be more appropriate where the technological change impinges on core activities, will be pervasive and cut across departmental boundaries and will have numerous users—as in digital transformations. However, even in minor applications, the existing power structure will be disturbed in some way by new technology. The digital technologies described in Chapter 16 also represent new organisational resources around which ownership struggles will cluster. In this way, the transformation process will reproduce, but also amplify, existing strains in the organisation's political system. In the last three years we have seen many automation projects—good projects in principle—die deaths, especially in risk averse cultures, for example insurance, because the change agentry has been seriously lacking.

In practice, political activity may well reach its highest throughout implementation as people's fears become realised. This can be a function of people having avoided facing up to the implications of IT, and now automation and digitalisation, at earlier stages. Also, failure to participate in change will be rational behaviour where the rewards from change do not seem commensurate with the efforts necessary, where there is a good chance of feeling manipulated, and where the cultural norms of the organisation do not encourage trust and openness in interpersonal relations. It also may be a function of the implementation approach. Historically, all too often new technologies have been implemented with the minimum of participation, the output of selective and seemingly reassuring information, and little regard for the labour implications.

In summary, it is crucial to understand a particular organisation's political structure and how different types and levels of IT, automation and digitalisation relate to political activity. Such understanding leads on to managing the 'shadow' politics track in order to maintain momentum for the digital transformation program. What do the executives responsible need to do?

The Power Audit

We suggest a preliminary step is carrying out a power audit that analyses the existing political and cultural system of the organisation. This means establishing membership of the dominant coalition and the sources of their influence. It also requires close analysis of lower-level participants, their power bases and the manner in which they, as well as the dominant coalition, are likely to support or challenge any level of transformation, and how crucial such responses might be to digital and organisational progress.

A second step is establishing the likely power bases of the change agents in the context of the existing political and cultural systems. Also, how these power bases can modify in the course of DT.

The third step is to establish a range of possible technology, process, strategy-structure, work restructuring and reward options, and their differing effects on the existing distribution of power. More radical shifts in power distribution will require different strategies and alliances, and a much more meticulously derived political approach.

The fourth step requires an assessment of the political and cultural problems likely to be engendered in the course of developing and implementing each of the feasible options identified. A preliminary assessment of the political feasibility of each proposed strategy and operational model can then be made.

The fifth step is to develop strategies and tactics to support the more likely options, together with an assessment of their probable success.

Mobilising/Gaining Power

An age-old question at any time, in any organisation. How can influence be managed? Political goals need to be formulated, and an ends-means analysis carried out (i.e., what do we wish to achieve, can we achieve it, what can we achieve, and how do we go about achieving it?). Targets to be influenced are then identified, and incentives desired by the targets determined. Change management involves mobilising these incentives and monitoring, then acting on the results.

The management of politics and culture continues throughout the DT program. In organisational settings, a political approach is rarely just a matter of accumulating enough power at the beginning of a project to do what you like thereafter. Various approaches to influencing the change process of are possible. We have suggested above, a structure that sees a senior executive 'visionary', a project/program champion as a senior 'fixer', project managers, and, of course, the program governance staff and constitution have a critical role to play. A steering committee including senior-level managers could become actively involved in the more political aspects of the DT program. Under this umbrella, those responsible for managing change can establish credibility and influence events tactically, for example:

- Make sure you have a contract for change.
- Seek out resistance and treat it as a signal to be responded to.
- Rely on face-to-face contracts.
- Become an insider and work hard to build personal credibility.
- Co-opt early those likely to be affected.

Tactics at this level are limitless. For example, we have seen (though do not necessarily approve of): presenting a non-threatening image; aligning with powerful others; developing liaisons; developing stature and credibility by attending to stakeholders' immediate needs before gaining approval for a less well understood project; and diffusing opposition by open discussion and bringing out conflicts. Certain power resources can be mobilised. A crucial one is expertise. Change agents can increase stakeholder dependence by heightening uncertainties to which the application of their expertise would be a remedy. Internal consultant activities across departmental boundaries may give privileged access to, and so control over, organisational information. Political sensitivity and establishing relations with those with power is also important, as is the gaining of 'assessed stature' by identifying and serving the interests of relevant others. Group support by departmental colleagues and related groups is a further power source that can be developed.

Implementation Problems

A political approach needs to address at least three execution challenges. One is lack of support for change. Another is the danger of the change process running out of control. The third, and related, problem is how to maintain influence over the political dynamics of change. It is important to bear in mind that digital transformation must be seen in the context of wider objectives than the limited goal of gaining acceptance for a specific system or technology. Technology and system modifications may be necessary in the face of politico-cultural difficulties. Implementation serves as an important period for learning about system design imperfections at the technical level as well.

Techniques for making change more supported include:

* making visible any organisational dissatisfaction with present processes, systems, ways of working.
* addressing people's attention to the consequences of not bringing in the technology and changes.
* building in rewards and reasons for people to support the transition and the future system.
* developing an appropriate degree, level, and type of participation for different affected parties.
* giving people the opportunity and time to disengage from the present state.
* setting up temporary structures to maintain control over the project in the transition period.

Also, adequate planning and resources help to see the transition through. There is also the need to develop and communicate clear images of the future to organisational members, and establish multiple and consistent leverage points, aimed not just at individuals but also at social relations, task and structural changes. It is also important to build in feedback mechanisms to monitor developments as early as possible.

Keeping power on the side of the DT program and handling the power dynamics of change is crucial. There are several activities that need to be kept in balance. Leaders and key groups will maintain active support for change, a culture and climate of success must be created, but also enough stability must remain. The pace of change has to be judged, so that the changes remain acceptable to involved parties.

Political tactics must not run out of control and must be consistent with and supportive of an overall change management strategy. Typically for DT programs, 'open' strategies involving widespread communication and participation work best where there is underlying support for the aims of the project, where there is a large number of affected parties, where there are differing views on how digital transformation can be achieved, where power to resist the changes is widely distributed, and where stakeholder involvement will provide vital information for program development. 'Closed', much less participative strategies will tend to be preferred where the benefits of participation are low, where there is widespread agreement and support for digital transformation, where its promoters are all powerful, or where the level of disagreement and hostility about the digital transformation is so high that participation is perceived to serve little purpose. However, the evidence is that in today's organisations many, but not all, 'closed' strategies are associated with disappointing results.

Conclusion

The evidence is clear: at the heart of change is a change of heart. But this is not about an overnight conversion like Saint Paul experienced on the road to Damascus in the Christian bible. Organisations require very strong foundations worked at painstakingly with senor executive commitment over long time scales, and navigated with the payoffs and staging points clearly in mind.

To bring home the relevance of change management framework and the complexity of the process, Chapter 22 focuses on an actual case of a company rewiring for digital transformation. This really is a case of managing beyond the quick fix. By 2023, DBS Bank Singapore had long become globally recognised as an exemplar of how to achieve digital transformation. DBS Bank (DBS) grew from a Singapore bank to become the largest bank (by assets) in Southeast Asia. It provides a full range of financial services for institutional banking, consumer banking and wealth management, and has over 280 branches in 18 countries, mainly in South East Asia, South Asia and Greater China. How was this achieved? As we shall see, this was through a recognised strategic imperative; distinctive digital leadership; building agile and scalable digital operations; designing new digitally enabled customer experiences; and incubating and accelerating new digital innovations. Chapter 22 brings together the themes of this section on digital transformation.

22

A Case in Point: DBS Bank

It's not a great thing to say, but the (2020–22) crisis has been good for DBS. We have leveraged the opportunity to put more daylight between us and our competitors.

DBS CEO, Piyush Gupta

Introduction

To bring home the relevance of the KCP change framework and the complexity of the process, let's examine an actual case of a company rewiring for digital transformation. By 2023 DBS Bank Singapore had long been globally recognised as an exemplar of how to achieve digital transformation. During the pandemic crisis and economic slowdown of 2020–2023, it still managed to make two potentially transformative acquisitions, launch two new digital exchanges and think anew about what future banking would look like. DBS Group achieved a record performance in 2022 as net profit grew 20 percent to S$8.19 billion. Total income rose 16 percent to S$16.5 billion, crossing S$16 billion for the first time. These were the fruits of long-term corporate dedication to becoming a digital business.

DBS Bank (DBS) grew from a Singapore bank to become the largest bank (by assets) in Southeast Asia. It provides a full range of financial services for institutional banking, consumer banking and wealth management, and has over 280 branches in 18 countries, including the priority markets of Mainland China, Hong Kong, Singapore, Taiwan, Indonesia, and India.

© The Author(s), under exclusive license to Springer Nature Switzerland AG 2024
L. P. Willcocks et al., *Maximizing Value with Automation and Digital Transformation*,
https://doi.org/10.1007/978-3-031-46569-7_22

Digital transformation has proven remarkably successful. In 2014 revenues reached S$9.62 billion and profits S$4.05 billion, both for the first time. DBS Group achieved another record performance in 2019 as net profit rose 14 percent to S$6.39 billion, while total income increased by 10 percent from 2018, to S$14.5 billion. Reporting in November 2020, DBS bank third-quarter net profit had increased 4 percent from the previous quarter to S$1.3 billion, while profit (before allowances) declined 9 percent to S$2.04 billion. Total income was 4 percent lower at S$3.58 billion. DBS bank was riding well the 2020–2021 pandemic and economic crisis. Its robust technology infrastructure has led to stable growth and DBS being recognised as one of the world's strongest and best-capitalised banks. In 2021 DBS were voted the world's best bank for the fourth year running. How was this achieved? Drawing on KCP research and composite sources, we describe the bank's evolution and distil out five key practices.

Practice 1—Vision and Strategy: Becoming the D in GANDALF

In 2009, DBS had the worst customer satisfaction scores of all the banks in Singapore. New arrival chief data and transformation officer Paul Cobban was embarrassed at the time to tell people who he worked for, because of the bank's poor reputation. Meanwhile DBS executives had seen rising demand from digitally savvy customers. Asia had over 700 million digital banking users in 2014, and this grew to exceed 1.9 billion by 2021. Younger, more mobile-centric customers preferred to engage with banks differently, the obvious example being smartphones. There were emerging fintech threats. There was also the question of how a Singapore-based bank could grow internationally. In 2009, after 27 years at Citigroup, Piyush Gupta was appointed CEO at DBS, and tasked with leveraging digital technology, not just as an infrastructure platform for growth, but also to accelerate the pace of banking innovation for increasingly tech-savvy Asian consumers. Speaking in 2010 he spelled out the vision:

> If we are able to leverage fintech, and offer banking through digital channels, the need for a large geographic footprint in order to scale up in large geographies such as China, India or Indonesia becomes less of an imperative. A successful digital banking strategy will enable us to accelerate our access to emerging markets without the need for a large and expensive bricks and mortar network.

DBS formed a strategy to put technology at the core of the business, spending some S$600 million annually between 2010 and 2014, then another S$200 million a year from 2015 to 2017. One notable aspect of strategy was the bank's determination to learn from hi-tech companies, rather than other banks, however digitally advanced. This came to be called GANDALF, standing for Google, Amazon, Netflix, DBS, Apple, LinkedIn and Facebook. DBS also learned from the Chinese equivalents of these businesses, in particular Ten Cent and Alibaba. GANDALF represented a set of principles—on culture, processes, data science, software—for tech teams within the bank to align the technologies they were developing.

Practice 2—Leadership and Management Development

At DBS, the CEO Pyush Gupta personally championed the digital agenda. He persistently asked how senior executives were exploiting the digitalisation of banking products and services. The Group Executive and Head of Group Technology Operations reported directly to him, and were on the central executive committee. Regular meetings identified digital disruption by competitors. Annual joint business and technology executive workshops pinpointed multiple technology investments. This drove strong business ownership of the technology roadmap. Senior business executives were encouraged to think digitally and be techno-entrepreneurs. Investments in collaborative technology brought lower-level managers into the change agenda, with, for example, an internal e-forum established in 2011 for open discussions about digital across the organisation. The main developments sound very familiar to those reviewing IT success stories over the years. Prior KCP research has found all the following as effective practices in major IT-enabled organisational change projects:

- A visionary CEO; CIO in senior executive team and reporting to CEO; convergence of reporting to CIO.
- IT investment prioritisation; enterprise governance and coordination; technology road mapping workshops.
- Integrated technology communications platform; enterprise portal; collaborative platform.
- Managerial mind-sets—IT and digital as 'first order' thinking among business executives: from 'tech idiots' to 'digital warriors'.

Practice 3—Agile and Scalable Digital Operations and Platform

On operations, the Group T&O Division was formed by merging Technology and Operations, creating an entity of over 5000 employees under one Division Head. The new T&O structure was closely matched with the business and geographical markets, with the T&O heads for each line of business reporting directly to Division Head, as did country T&O heads. The first objective was to rationalise and standardise the technology and operations platforms. But since the Asian markets were diverse and at different stages of development, the platform had to be both scalable and flexible. This led to acquiring configurable enterprise systems, modernisation of legacy systems, and the adoption of service-oriented architecture and enterprise application integration (EAI) technologies. Enterprise architecture needed careful development. Business and technology managers rationalised the banking applications needed to support the new business operating model. Deviations from the technology roadmap were carefully analysed and managed.

As a result, Finacle, the reworked and new core banking system, was deployed across 13 countries in 28 months. This facilitated the rapid rollout of new products, and greater flexibility for different country-specific preferences and product packaging features. Another enterprise-wide platform, introduced in 2011, was for wealth management; it integrated retail banking and private banking functionalities. This supported better management of customers whose wealth positions kept changing, and a more seamless customer experience. Customer could access and manage their accounts across the different DBS lines of business. DBS doubled its wealth management income from $505 million in 2010 to S$11 billion in 2014.

IDEAL—an Internet and mobile banking platform for businesses—followed in 2012. While region-specific, IDEAL was fully integrated with DBS systems, and also with clients' internal ERP systems. Using IDEAL, business clients could manage cash and trade financial transactions through straight through processing, as well access statements and foreign exchange functions. DBS purchased 'best-of-breed' systems but also built strong, responsive in-house competency in technology integration for devising reliable and relevant business solutions. In 2010, DBS had recruited a new head of process transformation to improve the bank's key processes, through process improvement methods and lean principles. The first year saw some S$6 million savings. The program was extended across the bank, resulting in numerous improvements, including by 2013, the elimination of over 240 million hours of customer waiting time.

Prior KCP research established the criticality of the organisation of the IT function, and of developing and aligning IT enterprise architecture and infrastructure with international business strategy. DBS bank corroborates these findings. It focused on:

- Restructuring the IT function by merging technology and operations (T&O), and aligning T&O with business divisions and geographical markets. Creating a head of process transformation and a process transformation office.
- Reengineering processes through streamlining, rationalisation, consolidation, improvement, enterprise architecting, while at the same time developing technology integration as a core organisational competency.
- Building the technology platform using digital technologies, including Internet banking, mobile banking, configurable core banking system, modernised legacy systems and enterprise application integration technologies.
- Focusing management and staff on lean thinking and continuous process improvement.

Practice 4—New Digitally Facilitated Customer Experiences

In 2010 DBS set up a Customer Experience Council, chaired by the CEO. The Head of Process Transformation also became Head of Customer Experience reporting to the Head of Group T&O. Given its prior poor record on customer satisfaction, the bank wanted a mindset shift beyond a culture of customer service. Designing improved customer experiences drew upon core Asian values that translated into: *"We are Respectful, we are Easy and Dependable to bank with (RED)"*. These core RED values were imbedded in attitudes, processes, and products.

A further focus was on better use of data to drive decisions in serving customers. For example, customer hours saved were formalised as strategic key performance indicators that management carefully tracked. T&O started to track and analyse customer pain points in their transactions with DBS. The resulting improvements saved 100 million customer hours. Another example. The 1100 ATMs in Singapore processed over 25 million transactions a month. Using analytics and sensors embedded in machine-to-machine (M2M) communication of its ATM network, DBS came to predict accurately ATM usage and withdrawal times. This led the bank to optimise its cash reloading schedule, cutting empty ATMs by over 90 percent, and reducing the number of customers delayed by cash reloading by 350,000 per annum. A customer touch point had transformed from an output channel to a customer real-time information sensor point.

In 2012 the bank also introduced speech analytics into its call centre that handled five million calls a year. This resulted in better identification and anticipation of customer needs and their complaints, and greater customer satisfaction. Also, a 5 percent reduction in time required to meet each user request. DBS also added alternative banking touch points as more transactions occurred outside traditional branches (e.g., self-service kiosks, and partnerships with pharmacies and supermarkets). In remodelled branches, bank staff were supported by a range of digital technologies and services that meant that 95 percent of interaction could be done without staff leaving their seats, while customers experienced faster, more seamless service.

By 2015 a customer could get notified a mobile-based queue number and wait time prior to a branch visit. As this freed up customers' time, they became more amenable to speaking with staff, resulting in double-digit sales growth for bank assurance and investment products at the branches.

Prior and ongoing KCP research establishes that automation and digital technologies are being increasingly leveraged for improving customer engagement, experience and establishing long-term relationships as a key area for competitive advantage. DBS has been a pioneer in this. Their practices over the years have included:

- Establishing a new head of customer experience, and a customer experience council and office.
- Developing and diffusing a customer service framework; operating human-centred, customer-centric design principles; creating a customer journey analysis lab.
- Building the technology for a new branch model, providing call centres with voice analytics, and developing data-driven customer analytics.
- Focusing employees on the total customer experience.

Practice 5—Continuous Digital Innovation

Our KCP research found four top recommendations for digital success: think, then act strategically, start right, institutionalise fast, and innovate continuously (see Chapter 1). DBS bank is consistent with these findings, including on the last.

In 2010 DBS set up an Innovation Council chaired by the CEO. Also, an Innovation Office. The bank also appointed a new Chief Innovation Officer reporting directly to the Group Head of T&O. Innovation training for staff, recruiting people with strong innovation potential, and crowdsourcing innovative ideas from employees quickly followed.

Many business initiatives resulted. An example was the uGOiGO on-line time deposit group-buy campaign in Hong Kong that targeted affluent customers using social media. Due to its success, this was quickly trade-marked and replicated across the region. Internally, mobile banking, digital payments, SME banking, and wealth management were amongst the targets identified for strategic digital innovation.

Singapore and Hong Kong were ideal test beds for 19 mobile apps. An example. Paylah, a mobile wallet for peer-to-peer fund transfers across DBS account holders. In Singapore this attracted 200,000 in six months. The mobile banking platform also grew to one million users by 2014. Twenty-four million transactions were being made via the Internet every month, and 11 million through the mobile apps.

DBS also worked with external partners to innovate further, for example with IBM, tech start-ups, and research institutions. This was also part of the drive to develop an innovation culture within DBS. Another approach was to run thousands of limited experiments across the bank to get people to behave more entrepreneurially and innovatively—as they would if working in a tech start-up.

Our ongoing KCP research establishes that, in their efforts to become digital businesses all too many organisations leave a massive amount of value on the table. The key, we find, is to create a digital options platform, as DBS has done, that widens massively the business options available, enabling innovations to proliferate, provided you have in place supportive innovation processes. DBS is representative of leading practices, including:

- Creating a new head of innovation, an innovation council and an innovation office.
- Providing new training, recruitment, assessment, promotion. Putting in pace strategic incubation, internal crowdsourcing, hackathons. Achieving innovation through technology partnering and start-up collaboration.
- Leveraging emerging technologies—e.g., mobile payments, analytics, social media. Learning from and deploying disruptive fintech practices and technologies, e.g., crowdfunding, P2P lending, social trading.
- Introducing and sustaining innovative, inventive, entrepreneurial, start-up like thinking.

Lessons: Digital Transformation and Competitiveness

DBS took a transformational approach consistent with the framework outlined in Fig. 21.1 (see Chapter 21). Notable throughout the DBS approach is the centrality of cultural change, and the recognition of the interrelatedness of all the components of culture, management mind-sets and skills, structures, reward systems, technology and processes, people skills and teaming. In particular, DBS developed interrelated capabilities in the areas of leadership, operations, customer needs, and innovation.

We note one additional factor implicit in our change framework that the DBS bank case makes transparent. It relates to the Navigation core capability described in Chapter 18. DBS is highly focused on metrics, taking a balanced scorecard approach. Metrics are designed to focus on direction, and ensure targets are achieved. At DBS traditional measures represent 40 percent of the balanced scorecard, while new areas of focus represent a further 40 percent. One of the four main areas of focus is: becoming a technology company. This involves embedding a Platform Operating Model, strengthening data infrastructure and implementing a robust data governance framework.

Distinctively, 20 percent of the metrics cover digital transformation ('Make Banking Joyful'). These measure progress in: growing meaningful relationships and outcomes with ecosystems partners; acquiring new customers and digital channel share; eliminating paper waste and driving for instant fulfilment; driving customer engagement and conversion across digital channels. Given that digital customers give higher income, better cost-income ratio, and higher return on equity versus traditional customers, DBS measures its progress in driving customer digital behaviour. DBS also measure its progress in embedding itself in the customer journey to deliver differentiated experiences.

Conclusion

Contrasted against how DBS went about digital transformation, it becomes more understandable why so many businesses struggle with becoming digital businesses. But pursuing digital strategy is today a key business imperative, however hard to deliver on.

To add even more urgency, it is worth reiterating that the evidence shows that being slow to adopt digital technologies may reduce risk in the short term, but builds growing business risk in the long term. Competitively, across sectors, top performers are gaining disproportionately large gains, with correspondingly heavy losses for those falling behind. The 'superstar' companies constitute the top 10 percent, and capture 80 percent of the economic profit. They come from all sectors and all regions of the world and are not limited to the obvious hi-tech US and Chinese firms. They include old global brands like Coca Cola and Nestle, but also a panoply of other firms, from Chinese banks, to French luxury companies to German automakers.

Early adoption of digital technologies has been helping to widen further this competitive gap, and this trend is being repeated and magnified by automation and 'AI'. According to some sources, early automation adopters, as front-runners, will gain 122 percent in cash flow between 2018–2030, while followers will gain only 10 percent and laggards will lose 23 percent.

All this makes even more vital the means to execute digital business strategy. The DBS bank approach is very hard to replicate, and therein, along with its continuous digital innovation, lies the bank's sustainable competitive advantage. However, it does raise questions, and provide lessons about urgency, sustained focus, long term investment, and capability building that all too many organisations, across sectors, need to heed if they are to become competitive digital businesses in the next five years.

Index

© The Editor(s) (if applicable) and The Author(s), under exclusive license to Springer Nature Switzerland AG 2024
L. P. Willcocks et al., *Maximizing Value with Automation and Digital Transformation*,
https://doi.org/10.1007/978-3-031-46569-7